THE HOT-BLOODED

ADRIAN J. DESMOND

THE HOT-BLOODED DINOSAURS

A REVOLUTION IN PALAEONTOLOGY

THE DIAL PRESS / JAMES WADE
NEW YORK 1976

First published in Great Britain by Blond & Briggs Ltd.

Copyright © 1975 by Adrian J. Desmond

Manufactured in the United States of America

First American printing

Library of Congress Cataloging in Publication Data
Desmond, Adrian J   1947–
  The hot-blooded dinosaurs.

  Bibliography: p.
  Includes index.
  1. Dinosauria.   2. Paleontology—History.
3. Body temperature.   I. Title.
QE862.D5D45   1976   568'.19   76-190
ISBN 0-8037-3755-6

Contents

1. The crown of creation

In the third year of the French Republic General Pichegru stood poised outside the gates of Maestricht, ready to lay siege. The Revolutionary army, sweeping north to liberate Holland in 1795, had found its way barred by the fortress on St Peter's Mountain near the small Dutch town. Next to the fort nestled a small château, the residence of a local ecclesiastic, Canon Godin, and as such a deserved target for republican hostilities. French artillery pounded the garrison, yet curiously spared the villa. The building was also the shrine of a priceless relic, and Pichegru's orders had been to preserve it at all costs. After the capitulation, the general commanded his troops to ransack the house and seize the relic for the Republic. But he was too late, during the night of the battle the wily canon had stolen his valuable antique away to the safety of the town.

The treasure that held up the French advance was a pair of formidable-looking fossil jaws, remnants of a colossal beast that had haunted the pre-Adamite world. News of the skull, armed with dagger-like teeth in its four-foot jaws, had spread quickly across Europe, and the fossil itself had gained a certain notoriety by the extraordinary controversies surrounding it wherever it went. No one was able to say with certainty what sort of antediluvian monster it was. Even the creature's emergence had been heralded by a bitter feud, and a costly legal battle to determine the rightful owner. It comes as little surprise that its presence was known to the French general. The importance of the fossil skull, however, rested not in its dramatic entry into a revolutionary Europe, but in the profound influence it exerted on palaeontologists over the next half century. It was itself to help effect a revolution in man's thinking about the primeval life of his planet: it would make the idea of extinction credible. On the other hand, it was to mislead scientists for many decades in their attempt to understand the nature of the huge saurians living in Mesozoic times.

Throughout the eighteenth century, the Maestricht chalk quarries were renowned for their wealth of fossil shells, particularly such curiosities as ammonites, belemnites and stony sea urchins. St Peter's Mountain was an especially rich prospecting area; a series of chalk quarries had been cut right into the heart of the mountain and then opened out into vast galleries with roofs supported on tall pillars like a Romanesque crypt. With the opening of new chambers and the enlarging of the old the whole mountain became honeycombed with subterranean caverns. The extraction of so much chalk inevitably resulted in the unearthing of many unusual fossils, and collectors would wander the underground galleries with torchlights in search of these antediluvian remains.

1. The subterranean caverns honeycombing St Peter's Mountain in Maestricht.

Then in 1770 the mountain yielded up its most spectacular possession: quarrymen working deep in the interior, five hundred yards from the main entrance, came upon the jaws of a truly monstrous animal encased in solid rock. One of the local collectors, Dr Hoffmann, a retired German military surgeon who had worked the region for many years to supply the Teyler Museum in Haarlem, was hastily sought for his expertise. Anxious to acquire the impressive jaws for himself, Hoffmann generously rewarded the workmen, who obligingly removed the rest of this valuable trophy for him. The block was hewn out intact and Hoffmann returned home with it. The Dutch anatomist Pieter Camper was summoned to identify the puzzling jaws. He supposed that they were the remains of an ancient whale, a natural assumption bearing in mind the size of the jaws and the marine shells in the surrounding rocks. His son, Adrien Camper, however, declared surprisingly that the beast was not a mammal at all but a monstrous marine lizard. But since lizards of this size do not inhabit the world today others thought it more probable that the jaws had belonged to a prehistoric crocodile.

Meanwhile, the controversy surrounding this immense skull had become the talk of the town and eventually reached the ears of Canon Godin, who owned the pastures above the quarry. Realising that a valuable asset had slipped through his hands, he demanded the prestigious fossil, invoking his feudal rights in an attempt to reclaim the pre-Adamite relic. Godin sued Hoffmann and won his case, thus acquiring the fossil jaws, whilst Hoffmann was left to pay the court expenses. Canon Godin mounted the relic in a glass shrine and placed it on show in his country house near the mountain for the curious to come and see.

8

2. The discovery of the first sizable pre-Adamite beast caused a sensation – and presented the scientific world with a riddle. Hoffman is seen conducting the operation in 1770 as quarryhands remove the mosasaur's imposing four-foot jaws.

Justice took several years to run its course and when at last it did Hoffmann was dead. As Maestricht fell to the republicans, the canon hid his treasure in the town. Pichegru retaliated the following morning by offering 600 bottles of best wine to anyone who would bring him the fossil. Troops combed the city and a dozen grenadiers carried the fossil jaws back in triumph to the general, and from here the antediluvian beast was taken as booty to Paris where it was installed in the Jardin des Plantes.[1] Here its scientific worth was established beyond doubt by France's leading anatomist, Baron Georges Cuvier.

As the Reign of Terror gripped the capital the young Cuvier was far removed in the Normandy countryside, collecting molluscs and studying the local fauna, while supporting himself with a minor clerical post; Cuvier was above all an administrator. His talents and researches did not pass unnoticed and in 1795, while still only 26, he was called to Paris to work alongside the evolutionist Jean Baptiste Lamarck in the Jardin des Plantes. In Paris, Cuvier's flair for organisation was at once apparent, and his desire to reduce everything to order, both in zoology and in public service, as a systematiser of fossil remains and, under Napoleon, as vice president of the Ministry of the Interior whose task was the reinstatement of a strong social order, dominated his life's work. The rule of law was a political imperative under the new regime, just as a strong natural order was the zoological imperative, and both fell compatibly within Cuvier's domain. As an outcome of his exhaustive study of fossil and living animals, especially large quadrupeds, Cuvier was finally convinced that whole animal races had disappeared from the surface of the earth. Extinction was a hoary problem; there were deeply entrenched metaphysical reasons for discounting it, and the fine evidence for an unequivocal demonstration had always gone lacking. In the late seventeenth century the English naturalist John Ray had toyed with the idea of extinction, but eventually returned to the conventional view that such creatures

9

as the spiral-shelled ammonites could still be living on the remote corners of the planet. (The discovery of the related nautilus in the South Pacific partly vindicated this approach: the nautilus is the sole survivor of a race of tentacled squid-like molluscs living in chambered and often decorative spiral shells.) Cuvier's predecessor in the eighteenth century, Le Comte de Buffon, at first suspected that the mastodon remains found on the banks of the Ohio River may have come from a lost species. He too reverted to the more traditional explanation, suggesting that the American mastodon skull was no more than a mixture of elephant's tusks and hippo's teeth. So the question remained open until Cuvier's day, and it is ironical that the overthrow of one of the most cherished notions since Aristotle's time should be the lot of a staunch, and by all accounts rather pompous, defender of law and order.

It was only a matter of months after Godin's fossil jaws arrived in Paris as a war trophy that Cuvier finished his work on fossil and living elephants, and was able to announce quite categorically that some elephant types had indeed died out.[2] Extinction had become reality. Cuvier's tool for demonstrating this was a rigorous comparative study of all the available specimens. His technique was to compare the corresponding structures of different animals and refer any dissimilarities between them to their different functions and ultimately different modes of life. The American mastodon and the Siberian mammoth, he declared, were not identical to the living elephant and since it was unlikely that they still survived on the earth it can only be concluded that they were extinct. The simplicity of the argument is misleading, for against it stood a powerful, if rarely explicitly formulated, counter-argument resting in large measure on faith. This inhibitory notion was the concept of the plenum: that is, that a bountiful God could only have populated the world with every conceivable type of organism as a sign of his omnipotence. These organisms could be arranged into an infinitely graded chain from the smallest microscopic creature to man. Extinction broke the chain by introducing gaps, a situation deemed by the pious an adverse reflection of the Creator's power.

Cuvier's detailed factual evidence in support of the mammoth's extinction finally shook man's faith in the plenum and removed the first link from the chain. It also paved the way for the acceptance of other prehistoric monsters no longer surviving. Whereas the mammoth had lived in the recent past, the 'great animal of Maestricht', as it came to be called, had lived in very remote times and was even less like present-day creatures. Cuvier concluded that the further back in time one looked, the less creatures resembled those alive today. This was brought home most forcibly by the discovery in Bavaria of a pigeon-sized pterodactyl in 1784, a beast recognised by Cuvier as possessing wings yet also being a reptile. The earth in Mesozoic times was populated by exotic and alien beings that would be quite unfamiliar today. Whatever sort of catastrophe had wiped out the mammoth in the recent past must also have struck earlier in the earth's history, reasoned Cuvier, taking its toll of the bizarre Mesozoic reptiles.

Cuvier examined the Maestricht skull and confirmed that it was, as Adrien Camper had suggested, a lizard related to the monitors of the tropics, albeit one

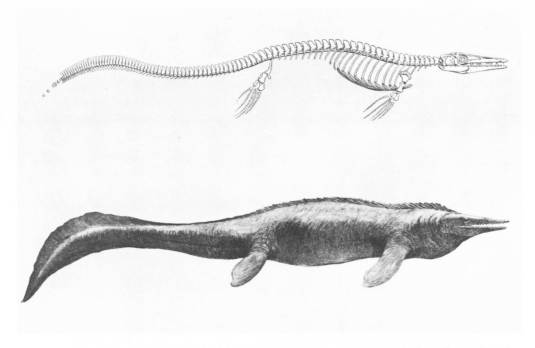

3. The *Mosasaurus*, an immense Cretaceous marine lizard, was to play a decisive role in the understanding of past life.

of stupendous proportions. Meanwhile, in England the Rev. William Conybeare, in the absence of any counter-suggestions, had at long last christened the creature: he called it the *Mosasaurus* or 'Meuse lizard' after the district from which it came.[3] Cuvier was entirely correct. But unlike the monitor lizards, the mosasaur was modified for a marine, fish-eating existence, with limbs turned into fins and a strong tail for propulsion. The Rev. William Buckland speculated that, from the length of the skull, 'this monstrous Monitor of the ancient deep was five and twenty feet in length',[4] even though the largest of its living relatives among the lizards barely reached half that size. When the first bones of what we now recognise to have been dinosaurs were unearthed, they were understandably assumed to have come from similar giant lizards. 'However strange it may appear,' said Buckland of the Maestricht mosasaur, 'to find its dimensions so much exceeding those of any existing Lizards, or to find marine genera in the order of Saurians, in which there exists at this time no species capable of living in the sea; it is scarcely less strange than the analogous deviations in the Megalosaurus and Iguanodon, which afford examples of still greater expansion of the type of the Monitor and Iguana, into colossal forms adapted to move upon the land.'[5] Buckland himself described the first of these terrestrial giants in 1824. Buckland was Professor of Mineralogy and Geology at Oxford University so the discovery of large bones in the Stonesfield Slate, only a few miles away, naturally attracted his attention and the fossils quickly found their way into the Oxford Museum.

Although no two of these bones were unearthed in contact with one another, with the exception of some vertebrae, Buckland was quite adamant as to the position of the creature in the zoological scheme of things, as the title of his paper 'Notice on the Megalosaurus or great Fossil Lizard of Stonesfield' suggests.[6] Not a little influenced by Baron Cuvier's convincing demonstration that the great Maestricht jaws had once belonged to a Mesozoic monitor, Buckland had naturally assumed the same of the huge vertebrae and limb bones in his museum. Indeed, the name that he and his accomplice in the catastrophist cause, the Rev. William Conybeare, concocted between them betrays the monster's supposed affinities: *Megalosaurus* means 'giant lizard'. In size it surpassed even Buckland's estimate for the mosasaur. The largest thigh bone in the museum was just less than three feet long and ten inches in circumference at its narrowest point. Informed of the existence of this bone, Baron Cuvier estimated that if the creature had the proportions of a lizard, it would have exceeded forty feet in length and have had the bulk of an elephant.

Like Cuvier, Buckland pointed to these reptilian remains as evidence that a catastrophe had laid waste to the land. But Buckland added a novel theological twist. Cuvier, even though a devout Christian, had been cautious of mixing Scripture and geology, and declined to speculate on the nature of the agent triggering the earthly revolutions. (He never questioned that they did have a natural explanation.) He also doubted the global extent of the revolutions; a drowning of individual continents would suffice. But when his *Preliminary Discourse* on those revolutions (first published in 1812) fell into the hands of its English translators in 1817, its tone was surreptitiously recast, with the Mosaic Deluge substituting for the latest revolution (thus lending Scripture a certain scientific respectability). It seemed to the English reader as though Cuvier himself was sanctioning the Flood. Buckland was heir to this English tradition. The Rev. Buckland, later to become Dean of Westminster, was a divine who shared none of Cuvier's reluctance to bring science to the aid of Revelation, and he so drastically modified the Cuvierian system that what for Cuvier had been a recent local marine incursion became a precipitous tidal wave dispatched by the Almighty to drown the globe. Buckland searched high and low – from the Himalayas to the caves of Yorkshire – for evidence that the Flood had been both universal and catastrophic. Moreover, just as the most recent extinctions had followed in the wake of a tidal wave of unimaginable force, so the Brobdignagian reptiles met their end in an earlier submersion. The Almighty, in a wanton act of 'Creative Interference', had swept away the old order; the Stonesfield fossil bones stood testimony to the periodicity of Heaven-sent global floods.

Gideon Mantell, a physician and fossil collector living in Lewes, not far from Brighton on the south coast of England, had found similar giant bones in Tilgate Forest. He too had a thigh bone of the *Megalosaurus*, but one that was twenty inches in circumference. Again, on the lizard analogy, this fossil was supposed to have come from a creature sixty or seventy feet long. Tilgate Forest was a rich prospecting area for giant saurians and it had already yielded remains of crocodiles and plesiosaurs to Mantell. He had also found enormous teeth unlike anything observed before. The crown of each tooth had been worn down into a smooth oblique surface and the shape of the tooth betrayed its owner as a plant-eater. The question of whether the teeth could possibly have come from a large reptile was arguable because there are so few reptiles today that can masticate plant material. Since Charles Lyell was about to visit Paris, Mantell gave him a tooth to show to Baron Cuvier, who immediately saw it as the upper incisor of a rhinoceros. Mantell could no more accept that rhinoceroses had lived alongside the giant reptiles than that the teeth had come from the overlying soil, as Cuvier suggested, so he took the fossils to the Hunterian Museum of the Royal College of Surgeons in London. It was there that he saw the teeth of the South American iguana lizard; the similarity was obvious and unmistakable. So alike were the small teeth of the iguana to the giant ones from the Mesozoic rocks that Mantell even toyed with the idea of giving both the same name! But upon the suggestion of Conybeare he decided to adopt the non-committal term *Iguanodon* or 'iguana tooth' for the Mesozoic forerunner.[7] If an iguana lizard had teeth this size, Mantell guessed that the whole creature would be upwards of sixty feet long.

Cuvier, Buckland and Mantell had all used monitor or iguana lizards as their models to gauge the size of the lizard's supposed ancestor on the ancient earth. They would scale-up the lizard so that its thigh bone, tooth or whatever was the same size as the fossil and then calculate the total size of the fossil animal. The fossil bones, especially the limb bones, were frequently fifteen or twenty times longer than those of a lizard, which led to fantastic dimensions for the extinct saurians. Buckland's description of a six-inch bone from the thumb of one of these ancient beasts found on the Isle of Wight leaves no doubt as to the awe these monsters inspired. 'It is,' he said, 'the largest metacarpal bone which has been as yet discovered; and if we apply to the extinct animal from which it was derived, the scale by which the ancients measured Hercules . . . we must conclude that the individual of whose body it formed a part, was the most gigantic of all quadrupeds that have ever trod upon the surface of our planet.'[8]

Cuvier's mosasaur lizard, the first of the Mesozoic leviathans to be unearthed, could scarcely have wielded more influence over the early palaeontologists. Many of the nineteenth-century fossil collectors were amateurs, and without their untiring spadework professional geologist-clerics like Buckland would have suffered greatly from a lack of raw material with which to work. Dedicated and persevering amateurs like Thomas Hawkins, Gideon Mantell and Mary Anning, living along the fossil-rich Mesozoic cliffs of England's south coast, were the first to disinter many of the awesome giant saurians that had once ruled the earth. It was these same visionaries, working the English Oolitic limestone (dating from the Middle Jurassic period) and the Lower Cretaceous sands and clays (from the weald counties of Kent and Sussex), who fired the public imagination with tales of sea monsters, flying dragons and huge land lizards inhabiting the earth before Noah. Most of the early pioneers were devout men and women and some appeared ecclesiastical fanatics even by contemporary standards. So a thorough investigation of the globe's archaic inhabitants not only lent an insight into the workings of Providence, but also rendered a distinct service to Biblical interpretation.

Because Cuvier had interpreted the first saurian as a giant lizard (which it was), *Iguanodon* and *Megalosaurus* were accorded a similar status. Cuvier was held in reverence and it was not considered wise to ignore his lead. Consequently, lizards had been the standard by which the extinct saurians were understood – and measured, a procedure which led to seemingly absurd sizes for the archaic reptiles. It was Richard Owen, Hunterian Professor to the Royal College of Surgeons in London, who finally broke with tradition. On August 2, 1841 Owen stood for two and a half hours delivering a 'Report on British Fossil Reptiles' to a meeting of the British Association convened that year at Plymouth. At 38, Owen was relatively young but the undisputed rising star of palaeontology, hailed by all as the English Cuvier. His broad knowledge of the Mesozoic fossil reptiles was second to none.

From among the saurians of Mesozoic times Owen singled out three species as being somewhat different from the rest. He announced that *Iguanodon*, *Megalosaurus* and Mantell's *Hylaeosaurus* (or 'Wealden Lizard', a colossal 'Fossil Lizard' bedecked with scaly armour plates, discovered by Mantell in Tilgate Forest in Sussex in 1832) were united by their unique possession of five fused vertebrae welded to the pelvic girdle. Unlike all other giant saurians, these were *terrestrial* rather than sea dwellers, and their huge limbs and stumpy toes gave them more than a passing resemblance to the heavy pachydermal mammals.[9] These characteristics, coupled with the enormous size of the beasts, were 'deemed sufficient ground for establishing a distinct tribe or sub-order of Saurian Reptiles,' proclaimed Owen, 'for which I would propose the name *Dinosauria*.'[10] Dinosaur means literally 'terrible lizard', a suitably evocative title given in recognition of the creature's dimensions and undoubted ferocity. But Owen had additional reasons for investing these giant reptiles with a little more importance than a formal study of their fragmentary remains warranted.

Unfortunately, history only honours those who fight on the winning side; Richard Owen was a party to the antiDarwinian cause in the evolutionary debate, for which history never really forgave him. Owen was caught at one of the major turning points in intellectual history. A brilliant practical anatomist, he was nevertheless a child of the established order and was unable to accept the changes that were in store. Owen entered a world where Cuvierian 'progressionism' was scientifically respectable: the evidence of time scattered races of extinct animals seemed to *demand* periodic catastrophes. On the other hand, the doctrine of evolution at this time was a romantic heresy, factually unsupportable and lacking any rigorous mechanism. Owen left a world where the fact of evolution was generally accepted, where the evidence had begun to shift in its favour, and where catastrophism was seen as a Creationist chimera. He was never able to adapt. As the tide of opinion turned after the publication of Darwin's *Origin of Species* in 1859 he assumed the role of chief spokesman for the antiDarwinist faction, relying on his prestige rather than reasoned argument to win his case. Time and again he suffered humiliating defeats, causing him to despise the evolutionists, of whom the brash young T. H. Huxley was the most vocal. Owen

– the English Cuvier – took his cause to the grave with him. History has of course been unfair. There *were* reasons in Owen's younger days for believing that evolution, especially as defined by its preDarwinian exponents, could receive no support from fossil studies. Dinosaurs were to be a crucial link in Owen's chain of reasoning.

Owen was sceptical of Buckland's method of calculating saurian dimensions using lizards as a standard. The resemblances of the *Iguanodon* to the iguana, he claimed with an air of authority, 'have misled the Palaeontologists who have hitherto published the results of their calculations of the size of the *Iguanodon*; and, hence, the dimensions of 100 feet in length arrived at by a comparison of the teeth and clavicle of the *Iguanodon* with the *Iguana*, of 75 feet from a similar comparison of their femora, and of 80 feet from that of the claw-bone, which, if founded upon the largest specimen from Horsham, instead of the one compared by Dr Mantell, would yield a result of upwards of 200 feet for the total length of the *Iguanodon*'.[11] Such dimensions were plainly nonsensical. Since this procedure was obviously suspect Owen instigated a new method. He measured the vertebrae and estimated their number from head to tail, thus arriving at the far more manageable length for *Iguanodon* of twenty-eight feet and for *Megalosaurus* thirty feet. These ancient saurians were clearly not as attenuated as lizards. Unlike the lithe and agile lizards, *Megalosaurus* was thought by Owen to be a massive, compact, and ponderous creature with long legs supporting its bulky frame.

> From the size and form of the ribs [continued Owen] it is evident that the trunk was broader and deeper in proportion than in modern Saurians, and it was doubtless raised from the ground upon extremities proportionally larger and especially longer, so that the general aspect of the living Megalosaur must have proportionally resembled that of the large terrestrial quadrupeds of the Mammalian class which now tread the earth, and the place of which seems to have been supplied in the oolitic ages by the great reptiles of the extinct Dinosaurian order.[12]

By 1841, when the dinosaurs were eventually recognised by Owen, the other major groups of extinct saurians had been known for some decades. Many were of an equally colossal size, which to a great extent lessened the impact of the dinosaur. Complete skeletons of ichthyosaurs ('fish lizards', streamlined marine reptiles shaped remarkably like dolphins, but lacking the dolphin's enlarged mammalian brain) and plesiosaurs (literally 'near lizards', swimming reptiles using paddles for propulsion) had been retrieved from the blue shales of the Lower Jurassic Lyme Regis cliffs on the Dorset coastline by Mary Anning and Thomas Hawkins, and Gideon Mantell had excavated the bones of giant crocodiles and mosasaurs in Sussex. So although dinosaurs were enormous in proportion to lizards, in Owen's eyes they were, in fact, hardly longer than the other leviathans of the archaic earth. Owen calculated that dinosaurs were little larger than elephants and rhinos, which they probably resembled in outward appearance; they were literally elephantine lizards.

If, then, size was of little criterion (although he often alluded to it to lend the group a measure of dramatic impact), why did Owen create the Dinosauria? Was it just that he wanted to give recognition to a few ill-defined characters, or was there some ulterior motive?

Owen's comparison of these reptilian giants to the elephants and rhinos provides the vital clue to his motivation. He believed that dinosaurs were the most advanced of all reptiles. 'The Megalosaurs and Iguanodons,' he claimed, 'rejoicing in these undeniably most perfect modifications of the Reptilian type, attained the greatest bulk, and must have played the most conspicuous parts, in their respective characters as devourers of animals and feeders upon vegetables, that this earth has ever witnessed in oviparous [egg-laying] and cold-blooded creatures.'[13] They were the crown of reptilian creation, a metaphor characterising Owen's prevailing beliefs with uncanny aptness. His researches on fossil reptiles had at last furnished the evidence that would make the evolutionary speculations of Jean Baptiste Lamarck untenable.

Lamarck had been Cuvier's colleague at the Paris Museum, but the two eminent men held diametrically opposed views on the history of life. Where Cuvier envisaged whole races of animals annihilated by mysterious catastrophes, and the emergence of entire new faunas to replace them, Lamarck saw a gradual succession of life, progressing through infinitely small changes. (Extinction, let alone mass annihilations, had no place in Lamarck's scheme of things.) Lamarck's evolutionary views had been prompted by the apparent increase in the complexity of animals on climbing the ladder of nature. 'It became therefore important to know,' asserted Lamarck in his *Philosophical Zoology* in 1809, explaining how he came by his evolutionary beliefs, 'how this organisation [of a simple creature], by some sort of change, had succeeded in giving rise to others less simple, and indeed to the gradually increasing complexity observed throughout the animal scale'.[14] There was an innate drive towards complexity, he assumed, so with each generation an infinitely small increase must occur; in time higher, more complex animals will evolve from simpler ones. This innate drive was assisted by the animal's increased use of certain structures; such use developed the structure in question (such as the giraffe's neck) and caused it to grow. Lamarck was treading on dangerous ground and Owen set about undermining the entire theory.

As a palaeontologist, Owen saw the weak point in Lamarck's arguments: if he could show that the increase in complexity was an illusion the whole edifice would collapse. Increasing complexity was a point insisted on on by the majority of early evolutionists and they thought it crucial to their case. 'For in ascending the animal scale,' insisted Lamarck, 'starting from the most imperfect animals, organisation gradually increases in complexity in an extremely remarkable manner.' This was the point Owen chose to demolish. Early evolutionists, before the appearance of Darwin's *Origin of Species* in 1859, by demanding an upward march of species from lower to higher forms, were bound to encounter resistance from the palaeontological lobby. Did the speculations of the evolutionists, asked Owen, derive any support 'from the facts already determined in the reptilian department of Palaeontology?'[15] If the present species of animals, argued Owen, have resulted from a progressive development of earlier species, 'each class ought now to present its typical characters under their highest organised conditions of organisation'. A glimpse at the fossil record shows this not to be the case. Today's lizards and crocodiles appeared to Owen lowly creations, hardly sharing in the magnificence of the Mesozoic dinosaurs. They are lower on the scale yet much later in time, a fact scarcely explicable by any sort of linear 'progression' of species. The superiority of the dinosaurs, living in a glorious 'Age of Reptiles' was a direct act of divine Creation. Species did not transmute into one another but were placed on the earth by Design and if they appeared to form a succession, it was a result of divine planning rather than evolution.

> The evidence . . . permits of no other conclusion than that the different species of Reptiles were suddenly introduced upon the earth's surface, although it demonstrates a certain systematic regularity in the order of their appearance.[16]

Despite the regularity, whereby the age of fishes gave way to the age of reptiles and that eventually to mammals, there is no ascent involved, but a plan of static organisation. The reptiles are not terminated at the present day by their most superior representatives, that is, 'at the Dinosaurian order, where we know that the Reptilian type of structure made the nearest approach to Mammals'. Owen's motive, therefore, for isolating the dinosaurs was to express this inherent superiority among reptiles. In so doing, he also armed himself with a means of attacking the evolutionists. Had there been any sort of ascent, as the pre-Darwinian evolutionists required, the pachydermal dinosaurs – the apotheosis of the reptilian cold-blooded grade – would be alive today. Instead, they lived on the ancient earth at a time when the Creator thought it fit for them.

The reason, according to Owen, for the Creator's choice of the middle period of earth history, at which time he populated the globe in abundance with giant reptiles of all kinds, was the dense and oxygen-deficient atmosphere. Reptiles are cold-blooded creatures; they require the sun's heat to warm them up as a prelude to activity. Their behaviour is geared to absorbing solar energy, rather than generating it internally by muscular activity like a mammal or bird. Consequently, a mammal can remain active regardless of ambient temperature, but it requires a much greater intake of oxygen to release the energy to sustain this protracted period of activity. The less efficient heart and lungs in a reptile oblige it to survive on less energy. In Mesozoic times, thought Owen, the extraordinary prevalence of giant cold-blooded creatures can only imply that there was less oxygen available. And since the reptiles took to the air as pterodactyls, presumably the atmosphere was far denser then to support them as floaters rather than flappers (Owen doubted seriously whether the pterodactyl's lowly physiology could have sustained energetic flapping).

The Mesozoic world designed by providence had admirably suited the cold-blooded reptiles that inhabited it, but was totally unsuitable for creatures of a higher organisation. It was only when the oxygen in the air increased and the pressure dropped that 'the beautiful adaptation of the structure of birds to a medium thus rendered both lighter and more invigorating' could come about. And the discovery of bones, apparently of birds, in the Wealden strata of the Cretaceous (actually the remains of pterosaurs) indicates the beginning of the process of 'invigoration' at this time. Its continuation left the world uninhabitable for the huge saurians, which died *en masse* at the end of the Cretaceous. But the *Iguanodon* and *Hylaeosaurus* dinosaurs were also found in the Wealden, so it was hardly surprising to Owen that they should show the closest approach in a reptile to the warm-blooded classes. Owen presumed that, like crocodiles, dinosaurs had a good circulation to cope with the more invigorating climate in which they found themselves. They needed a good four-chambered heart, with a complete division between the spent venous blood and the oxygen-rich arterial blood. In the lowly lizards the venous blood mixes with the arterial in the heart as a consequence of which the tissues receive only partly renewed blood and have to make do with far less oxygen.

> The Dinosaurs, having the same thoracic structure as the Crocodiles, may be concluded to have possessed a four-chambered heart; and, from their superior adaptation to terrestrial life, to have enjoyed the function of such a highly-organised centre of circulation in a degree more nearly approaching that which now characterises the warm-blooded Vertebrata.[17]

This is how Richard Owen concluded his paper – the very first paper – on dinosaurs. These were bold conjectures on his part and it was perhaps a measure of Owen's prestige and undisputed command of his subject that he was able to put forward such suggestions. 'A too cautious observer would, perhaps, have shrunk from such speculations,' he declared. Yet Owen's reasoning does not stand any degree of scrutiny. Why did an 'invigorating' climate render the world unhabitable for dinosaurs? Why did lizards survive? Owen obviously felt it necessary to go to such lengths to counter the theories of the evolutionists. Because they thought in terms of such qualitative attributes as 'progression' and 'higher' or 'lower' life forms, Owen was forced to do the same. As a consequence his arguments were as fraught with glaring inconsistencies as their own.

How are we to assess Owen's beliefs? We must not be misled into believing that he in any way anticipated modern ideas, based as they are on detailed anatomical and ecological studies. To combat the heretical speculations of the evolutionists he had been forced to create a totally new order of reptiles from only three known species and then to recognise them as the height of reptilian achievement. The dinosaurs, it seems, were created less by God than by human ingenuity. Owen literally populated his Mesozoic earth with dinosaurs as a ploy against the evolutionists. His train of ideas was formulated as a reaction. He *needed* a race of super-reptiles, so he created one.

Even though dinosaurs shared certain mammalian features, Owen certainly never swayed from his belief that these 'gigantic Crocodile-lizards' were, in fact, anything other than cold-blooded. It was this model of the dinosaur – a gigantic lizard – rather than his fanciful speculations (which were quickly forgotten once the evolutionists stopped clinging to the simplistic upward-striving linear progression of life) that had a lasting influence on palaeontologists. The dinosaur was pictured as a monstrous lizard, blown up to look like a scaly rhinoceros.

Owen's legacy was established in more concrete terms when, thirteen years later, he had the opportunity to build life-size models of his dinosaurs. The occasion presented itself when the central building housing the Great Exhibition of 1851, Joseph Paxton's Crystal Palace, the largest iron-framed glass building ever constructed, was moved from Hyde Park to a new location in the suburbs at Sydenham. The Hyde Park exhibition had been in celebration of the fruits of industrialisation and was designed as an international trade fair for technological manufactures. When it closed in 1852, the Palace was dismantled piece by piece and re-erected in Sydenham, where it was to form a permanent showcase for the arts and sciences.

It seems that it was Victoria's consort, Prince Albert, who first suggested that the landscaped grounds of the new park should be adorned with restored beasts of by-gone ages. The task of reconstruction fell upon a wildlife painter and sculptor, Benjamin Waterhouse Hawkins, who had functioned in a supervisory capacity at the Great Exhibition. Hawkins was given a *carte blanche* and he chose the building of large mammals like the mastodon. Then he came upon Owen's monographs on the giant reptiles of the Mesozoic and resolved instead to restore these creatures to their former glory. For the Herculean task of 'revivifying the ancient world' Hawkins worked closely with Owen from the planning to the modelling itself.[18] From miniature scale models no more than a few inches long, Hawkins constructed life-size clay moulds, the largest of which weighed thirty tons, and from these took his casts. Three islands were built to house the monsters on an artificially created six-acre lake in the Palace grounds. Real vegetation gave the islands a lush tropical setting, a scene enhanced by the construction of replicas of Cretaceous cyad trees – palm ferns with leaf-scarred trunks resembling petrified pineapples. To add to the illusion of a tropical beach (the creatures were assumed to have been coast dwellers) the waters of the lake were intended to rise and fall a few feet like an actual tide, partially submerging the creatures at intervals.[19] How this could have been achieved in a six-acre lake defies analysis.

The creatures represented the reptilian life peculiar to England in Mesozoic times, although there were some accompanying Tertiary mammals like the giant elk. Hawkins described his vocation as summoning 'from the abyss of time and from the depths of the earth, those vast forms and gigantic beasts which the Almighty Creator designed with fitness to inhabit and precede us in possession of this part of the earth called Great Britain'. The reptiles on the three islands form a sequence, with the archaic creatures of Triassic age at one end and the culminating dinosaurs guarding the younger Cretaceous end. Because Hawkins had only the skulls of the Triassic beasts, he was forced to resort to a certain amount of ingenuity in adding the bodies. He had only modern reptiles as a guide so naturally they provided his models. The *Dicynodon* or 'two tusker' was cousin to those reptiles that evolved into mammals. Since its skull had a horny beak

4. Waterhouse Hawkins' Crystal Palace studio as it must have appeared to Queen Victoria on her visit in 1853. The central figure is the *Iguanodon*, flanked by *Hylaeosaurus* (right) and the Tertiary mammal *Anoplotherium* (left). In the foreground squats the dicynodont (right), thinly disguised as a turtle, and the labyrinthodont amphibian, restored in the likeness of a giant frog (left).

resembling a tortoise's, Hawkins restored a huge tusked tortoise. The *Labyrin-thodon*, which derived its name from the labyrinthine folding of its tooth enamel, was a primitive armoured amphibian looking not unlike a crocodile in life. Hawkins restored three of these as giant frogs and toads complete with stumpy tails.

The dinosaurs were represented by Owen's three species (these were still the only dinosaurs recognised in 1854), the Cretaceous *Iguanodon* and *Hylaeosaurus* and the older Jurassic *Megalosaurus*. They appear on all fours looking like rhinoceroses (a similarity made all the more striking by the horn on the *Iguanodon's* nose – this was, in fact, the thumb bone). The mosasaur was restored lurking in the water with only its head breaking surface; in fact, only the head *was* sculpted, the body was still unknown even though Mantell had detected fragments in the English chalk. From a commanding cliff top vantage point two giant pterodactyls gaze majestically over the scene, whilst long-snouted marine crocodiles, ichthyosaurs and plesiosaurs bask on the banks of the lake.[20]

With the project completed Hawkins and Owen celebrated by holding a dinner *inside* the *Iguanodon* on the eve of the New Year in 1853. Twenty-one dignitaries occupied the reptile's socially loaded stomach with Professor Owen sitting at the head (literally). The group regaled in sumptuous fashion and the revelries went on into the night as a succession of congratulatory toasts were proposed. When the banquet finished in the early hours the whole party of eminent scientists could be seen climbing out of the dinosaur and staggering across the park whilst singing all the while the merits of Hawkins' restorations![21] 'Potentates of the Wealden and the Oolite!' cried an incredulous *Quarterly Review*. 'Saurians and Pterodactyles all! dreamed ye ever, in your ancient festivities, of a race to come, dwelling above your tombs . . . dining on your ghosts, called from the deep by their sorcerers?'[22]

Queen Victoria and Prince Albert reopened the Palace on June 10, 1854, in the presence of forty thousand spectators. The consensus of opinion among the public was clearly that these were antediluvian monsters, beasts that had been destroyed in the Biblical deluge, even though such a view had long been out of vogue in scientific circles. Even the literary journals, deliberating on this 'mausoleum of the memory of ruined worlds', as the *Quarterly Review* described the assemblage, argued the various merits of the Creator's past attempts. Spectators flocked to the Palace in their droves to witness the monsters that inhabited the earth before Noah, leaving a trail of devastation in their quest for souvenirs.

The Crystal Palace building was completely gutted by fire in 1936, which probably accounts for the dinosaurs' fall into obscurity. Today the reptiles can still be seen on their prehistoric islands, standing testimony to the Victorians' conception of the dinosaur, a conception that has dominated both the public and scientific mind ever since. Charles Dickens summed it up in the opening passages of *Bleak House*, conceived in the year that Hawkins began the Palace recon-structions. 'Implacable November weather,' wrote Dickens. 'As much mud in the streets as if the waters had but newly retired from the face of the earth, and it

5. A contemporary illustration of the dinner held inside the *Iguanodon* model, with Owen proposing the toast.

would not be wonderful to meet a Megalosaurus, forty feet long or so, waddling like an elephantine lizard up Holborn Hill.'[23]

So the early Victorians bequeathed to us the dinosaur as a monstrous lizard: a creature with all the connotations that are implied by 'reptile'. The Crystal Palace restorations have survived into the later twentieth century as a quaint piece of Victoriana, standing testimony to a belief that dinosaurs were, in Dickens' words, 'elephantine lizards'. With this more visual reminder has come a way of viewing dinosaurs as essentially cold-blooded creatures. Cuvier, in correctly recognising *Mosasaurus* as a lizard, was also unwittingly casting its dinosaurian contemporaries in the same mould. Owen may have emancipated the dinosaur from the ranks of the lizards, but he was unable (if, indeed, he was genuinely willing) to shake off the lizard-like physiology from the dinosaur. For over a century this picture of the dinosaur has remained unchallenged. The situation as it still existed in 1968 was reported by Robert T. Bakker, then at Yale University:

> Generally, palaeontologists have assumed that in the everyday details of life, dinosaurs were merely overgrown alligators or lizards. Crocodilians and lizards spend much of their time in inactivity, sunning themselves on a convenient rock or log, and, compared to modern mammals, most modern reptiles are slow and sluggish. Hence the usual reconstruction of a dinosaur such as *Brontosaurus* is as a mountain of scaly flesh which moved around only slowly and infrequently.[24]

22

6. *Iguanodon* as it appears today. Owen restored it after the fashion of a huge quadrupedal mammal.

7. The *Iguanodon* (above) and *Megalosaurus* (below) on their island at Crystal Palace.

24

Yet there are profound differences between the dinosaur and lizard, differences that lead us to assume that all is not well with our model of the dinosaur. Even a cold-blooded 'elephantine lizard', by the very nature of its size, would be a radically different beast from today's small lizards.

We have been talking about the lizard as 'cold-blooded' but this is a misnomer since its blood is not by any means cold. The distinguishing features of reptiles, seen in contrast to mammals and birds, is that they do not depend upon heat produced internally. The muscles and organs of the body in 'warm-blooded' creatures produce enough heat to maintain the body at a high stable temperature, a process that is controlled by the hypothalamus in the brain. Lizards, however, rely on the heat of the sun and the surrounding air to warm up their bodies. For this reason they are better called ectotherms: dependent upon external heat sources to raise their body temperature, in contrast to the endothermic birds and mammals.

Before lizards were studied in their natural environment, it was thought that their temperature varied directly with that of the air. At the turn of the century, physiologists monitoring lizards under laboratory conditions naturally found this to be the case. But in the wild lizards are able to raise their temperature *above* that of the air and then hold it within strict limits despite minor fluctuations in ambient temperature. The greater earless lizard of the southwestern United States, for example, can keep its body to within 3.3°F of its average of 101.3°F for 75% of its active life.[25] Most lizards, in fact, possess a high optimum temperature which is strictly maintained during their waking life.

Why were the laboratory observations so misleading? 'For much the same reason,' mused Charles M. Bogert of the American Museum of Natural History, 'that a man with a heavy iron ball chained to one leg cannot demonstrate how fast he can run!'[26] The lizard is able to maintain a high internal temperature by sun-orientated behavioural responses; confining it to a cage merely serves to inhibit its free range of activities. Most lizards are 'sun baskers' and exploit the sun's rays to the full by adopting a series of appropriate postures and orientations. In this way the lizard can vary its heat uptake quite precisely. If, for example, its body temperature drops below optimum it aligns itself at right angles to the sun's rays to maximise exposure. Similarly, in early morning or late evening when the sun is low in the sky the lizard will cling to rock faces and slopes so that its body receives the warm rays perpendicularly along its entire length to make the most of the available energy. If the lizard becomes too hot whilst foraging during the day it will turn and face the sun to minimise the area of exposure.

In this way some lizards are able to maintain a temperature that is as high as man's own. Indeed, some can stabilise much higher: an active American whip tailed lizard maintains its body at between 104° and 106°F. A high internal temperature is required for the optimum functioning of the body's biochemical processes; when the temperature drops the chemical reactions are slow to work and the reptile becomes sluggish. Each species has its own optimum temperature so that two species living side by side in the wild (lizards that would have the same temperature forced on them in the laboratory) can vary by as much as 12°F by adopting different responses to the sun.

Theoretically, the bulkier an animal the greater its volume relative to surface area and the longer the sun exposure time it needs to warm up. This, anyway, is the case for inanimate objects – but they are never quite as versatile as lizards. The relationship breaks down for small ectothermic reptiles precisely because of their marvellous heliotropic reactions. Larger lizards *should* heat up and cool down more slowly than their smaller cousins because of the relatively smaller surface area yet larger bulk, but in fact there is little difference because they adopt all manner of positions to maximise their exposure to the sun, whilst at the same time darkening their skin to make it more absorptive.

This is all very well when dealing with small creatures which take only a short time to reach their optimum temperature. There *is*, however, an upper limit in size beyond which these behavioural responses and colour change no longer enable the creature to compete with its smaller relatives. It would be totally impossible for an alligator to warm up as speedily as a small lizard. In an attempt to get a little closer to the dinosaurs (both in size and relationship) Bogert joined forces with dinosaur expert Edwin H. Colbert, also of the American Museum of Natural History, and Raymond B. Cowles of the University of California in 1946 and moved on to the alligator as a test animal. At the Archbold Biological Station, near Lake Placid in Florida, the American alligator could be studied in its natural environment, one of palms, cycads and coniferous trees that may have approximated closely to the dinosaur's own Cretaceous world. The team worked with thirteen alligators, ranging from one foot to almost seven feet long and weighing from a few ounces to fifty pounds. The animals were tethered to stakes in the sun and their temperatures recorded over ten-minute intervals. In the initial experiments – as expected – it was immediately apparent that in direct sunlight the rate of temperature increase of the alligators varied inversely with size. The smallest alligator required only ninety seconds to raise its temperature 1° whereas a thirty pounder needed five times as long. 'Continuing this line of reasoning,' said Colbert, Cowles and Bogert, 'it would seem probable that in an adult ten ton dinosaur, say an animal with a body weight of about 9 million grams, the rate of temperature rise would be very much slower than a large alligator. Indeed, if the same difference in temperature rise as existed between the large and small alligators were applied to the dinosaur (an animal 700 times greater in body mass than the *large* alligator) then one may suppose that it would have taken more than 86 hours to raise the body temperature by 1°C in the adult extinct giant.'[27] In other words, if the dinosaur's temperature dropped just one degree below its lower threshold for activity, it would have to bask in the blazing sun *for over three days* to bring its temperature up to normal! (Assuming no night time intervened of course.) This is absurd. What it does mean is that *assuming* that there was a constant high ambient temperature then the dinosaur's immense bulk was capable of retaining the heat overnight so it started a new day with barely a drop in its heat reserves. If Colbert's figures were correct, dinosaurs could have held their temperature absolutely stable (within a few degrees leeway of optimum) for all of their adult lives. *If*, that is, they inhabited regions with high stable temperatures. Although they were 'cold-blooded', they never became cold. This seminal work on alligators established that the elephantine nature of the 'lizard' bestowed upon it a temperature pattern like that of a mammal or bird even though its physiology and metabolism were supposed to be totally distinct.

There were numerous objections raised by co-workers to this excessive figure of 86 hours for a one degree rise in temperature, so in the following year Colbert, Cowles and Bogert lowered their estimate to 'several hours', whilst arguing that unless a much larger crocodile were used as a test animal the exact rate of heating and cooling in a dinosaur would never be known.[28] By the time Colbert came to write his *Dinosaurs* in 1962 he had dropped his estimate to two or three hours for a 1° rise in a thirty ton *Brontosaurus*.[29] Bogert lowered the figure still more: a ten ton dinosaur, he argued, would have to bask in the blazing sun for a little more than an hour to elevate its body temperature by two degrees.[30] The lowering of the estimate changed the situation drastically. If we accept the revised figure, the ectothermic dinosaur *could* have lost an appreciable quantity of heat overnight, as Bogert foresaw:

> Suppose our hypothetical dinosaur were active in daytime, and subject to cooling at night, as it would be in any desert region today. If its temperature dropped even four or five degrees below its threshold for activity, the dinosaur would have to bask for a large part of the next day in order to regain the threshold temperature of activity.[31]

Basking in the hot sun for an uninterrupted number of hours would have had grave consequences for the dinosaur. Colbert's work on the alligator fore-shadowed these consequences rather dramatically. He measured the *internal* temperature of his alligators, but when that reached a near optimum two of his subjects suddenly died. The reason was that the sun was blazing down on the outside of the animals' bodies and the heat was having to be carried to the tissues and organs by conduction and transport by the blood. Because of this a gradient was built up so that when the temperature had reached only 100°F internally the armoured surface was so hot that the animals perished. Colbert extrapolated the time spent heating up from the alligator to the dinosaur but he neglected to extrapolate the heat gradient between the inner and outer surface. Giant dinosaurs were characterised by immense bulk but relatively small surface area and the only way the sun could heat up this inner mass of flesh was by beating down on the exposed dorsal surface. A dinosaur, like a lizard or mammal, operated at an optimum temperature (whatever it may have been). Yet to raise that temperature in the very depths of its body, the skin would have to have suffered a blistering heat which would tail away until the optimum was reached internally. If the gradient was considerable in Colbert's Mississippi alligators, it must have been simply enormous in fifty-ton dinosaurs. If the blood failed to dissipate the surface heat quickly (by carrying it away) in the tethered alligator, it would surely have failed to do so in a dinosaur. It therefore seems highly improbable that a fifty-ton dinosaur could ever have achieved a uniform optimum temperature by basking in the sun like a lizard.

This is something of an oversimplification. But it does illustrate how this study in the mid-1940s raised far more questions that it could answer. Colbert's alligators were tethered in the sun. Had they been resting on a river bank they would have been obliged to retreat into the water or shade long before reaching a critical condition. So an untethered ectothermic dinosaur would have had to have spent its time switching between sun and shade, trying to strike a balance between its internal and surface temperatures. Assuming, that is, that it did not *have* to stand out in the blazing sun all day to make up the heat lost the previous night. Since the amount of heat lost overnight and the length of exposure time needed to replace it are unknown it is difficult to know whether or not a cold-blooded dinosaur would work. The considerable divergence of opinion on the length of the basking time necessary to raise the animal's temperature just one degree illustrates the speculative nature of the whole exercise. Precise data was lacking so it was possible, in the 1940s, to conclude that these reptilian leviathans were ectothermic (that was never questioned) but that they frequented tropical climes where there was never a severe temperature drop at night. Since dinosaurs roamed into the northerly regions far from today's tropics it was taken for granted that in Mesozoic times the earth was uniformly warm over a large part of its surface. Nevertheless, it is difficult to imagine a cold-blooded creature, whose tonnage exceeded that of a large family of elephants, keeping a stable high temperature by sunning itself: the contradictions inherent in a basking fifty-ton ectotherm seem insurmountable. The questions raised by Colbert's study were never answered; the sheer quantity of unknowns made the problem intractable. When two decades later the subject was looked at anew it was from a totally different angle, and the results of this new approach showed why it was that so many difficulties had been encountered using the lizard as our substitute dinosaur.

2. The tyrant finds its feet

Far from the urbane setting of Prince Albert's Crystal Palace festivities, the New World was experiencing its first dinosaur discoveries. Only here the dinosaur was to emerge in a bizarre guise totally undreamed-of by Owen. In 1854 a pathfinding expedition of the United States Topographical Engineers trekked west to reach the Judith River, near its union with the mighty Missouri, in what was then the Nebraska Territory (now Montana). Being an official scientific exploratory survey, the Government dispatched Ferdinand Vandiveer Hayden as the party's geologist with the task of mapping and gathering specimens. Hayden, though only 26, was a veteran of an earlier fossil hunting expedition, when he made his way west from New York to the badlands of the upper Missouri. Familar with badland terrain, Hayden scoured the Judith's desolate and rugged outcrops. Such exposed, weather-worn terrain, scarred by deep ravines and canyons and lacking shrub covering, made the pickings easy and Hayden returned east with a cache of large teeth taken from the Cretaceous sands and clays.[1] In early 1858 he donated his geological finds to the Academy of Natural Sciences at Philadelphia where Joseph Leidy, otherwise Professor of Anatomy at the University of Pennsylvania, was Director. Leidy's immediate reaction was to announce that Europe was not alone in having been populated by dinosaurs in the Mesozoic; the primeval American landscape too must have witnessed the rise and fall of these gigantic creatures. Some of the unworn crowns of the Judith teeth were conical and slightly curved, with a cutting edge characteristic of a plant-eating creature: hence Leidy designated this herbivorous saurian *Trachodon* or 'rugged tooth'. Just as Mantell had found the megalosaur a 'fit carnivorous contemporary' to prey upon his *Iguanodon*, so in western America in Cretaceous times there was a huge predator stalking the *Trachodon*. Hayden's expedition had brought back a dozen fangs of this carnivore. The long teeth, curved and serrated, must have been borne by a fearsomely-armed reptile, which Leidy evocatively titled *Deinodon horridus*, or 'most horrible of the terror teeth'. The pattern of wear on the teeth, Leidy noted, gave the impression that 'the upper and lower jaws . . . closed upon one another like the blades of scissors, so that they were well adapted for penetrating, tearing and cutting their animal food'.[2] While the placid herbivores *Iguanadon* and *Trachodon* quietly grazed along the shoreline of the Cretaceous inland water courses, the *Megalosaurus* and *Deinodon* infested the waters like crocodiles, ready to seize their contemporaries. The struggle for survival in the remote past so forcefully stressed by Leidy, with giant herbivores kept in check by giant carnivores, was all the more remarkable a scene for being based on only a few broken teeth. But by the 1850s Leidy had the experiences of the English fossil collectors to draw on. As in England, the dinosaur's teeth had emerged before its skeleton to give a tantalising glimpse of life in the distant Mesozoic. The discovery of the rest of the giant's remains was awaited with high expectancy, but when it came it delivered a severe blow to conventional ideas. When Leidy first saw a skeleton of one of his American dinosaurs, it did not coincide even remotely with Richard Owen's conception, which by now had become world-famous through the concrete effigies at the Crystal Palace.

Leidy did not have to await the arrival of the next western expedition for his skeleton. William Parker Foulke, a fellow member of the Philadelphia Academy of Sciences and regular contributor on geological matters, had spent the summer and autumn of 1858 at Haddonfield, New Jersey. He learned that a neighbouring farmer had unearthed large bones while digging marl (a type of muddy limestone) on his farm some twenty years earlier. Because of their size, these vertebrae had attracted a great deal of attention and had all been carried away as souvenirs. Since neither the head nor limbs had been dug out of the marl, Foulke set out to find the old quarry on the farmer's land. Haddonfield stood on Cretaceous rocks, so Foulke was hopeful of finding saurian remains, especially since mosasaur fragments had been found in the vicinity. The marl pit was eventually rediscovered by one of the original workmen in the bed of a ravine and Foulke hired experienced marl diggers to begin operations anew. Throwing aside the superficial layers of earth they encountered a thin layer of crumbling fossil mollusc shells at a depth of eight feet. Excavating still further, at ten feet they hit upon a hoard of ebony black saurian bones. These proved extremely friable and Foulke lavished a great deal of care easing out each one individually using only a trowel and knife. In case they should fracture on removal, a sketch was made of the bones *in situ* so they could be pieced together later.

At the Academy Leidy was informed and promptly visited the scene of the excitement, helping to remove the last remaining bones. Alerted to this treasure on his very doorstep, Leidy made a tour of all the marl pits in the area, enquiring after giant saurian remains, but though many workmen had seen and handled them, the bones had always been thrown away and lost! Leidy came away with a few crocodile and mosasaur fragments, but nothing to compete with Foulke's impressive skeleton. Foulke presented his fossil bones to the Academy on December 14, 1858 and Leidy set about their formal description. The remains were those of a large herbivore related to the English *Iguanodon* and Leidy composed the name *Hadrosaurus foulkii* to commemorate the creature's discoverer. Raking through the marl, Foulke had collected in addition nine shield-shaped teeth, all about two inches long and coming from the lower jaw – teeth that clearly allied the hadrosaur with *Iguanodon* and *Trachodon*. As a result of Foulke's painstaking coaxing, the brittle bones were retrieved without damage from the rock. Besides vertebrae, Leidy had the humerus, radius and ulna of the arm as well as the corresponding leg bones. Since these were found together it was certain that they came from one individual.

To everyone's complete surprise, there was a peculiar disproportion between the front and hind legs. The thigh bone was complete and measured forty inches, yet in the upper arm the humerus was a minimal twenty-three inches. Leidy freely admitted that had he not come across the bones together he would have made two quite different animals out of them: one known only from hind limbs and the other from front limbs! Estimating the number of vertebrae to have been about eighty, Leidy guessed the total length of *Hadrosaurus* to be twenty-five feet. Since the hind legs raised the rear of the reptile a clear six feet off the ground, while the front legs were under four feet long, the saurian in life could have borne no resemblance to the London *Iguanodon* model. And yet the two were closely related; Owen had taken more for granted than anybody had appreciated.

Supplanting Owen's rhinocerine *Iguanodon* brought its own problems: with what could it be replaced? What *did* the enigmatic dinosaur look like in life? There are very few modern reptiles whose hind limbs far exceed their front ones in size. Leidy's first impression was that *Hadrosaurus* stood and walked like a kangaroo, even though this was a mammal and not a reptile. 'The great disproportion of size,' he wrote, 'between the fore and back parts of the skeleton of *Hadrosaurus*, leads me to suspect that this great herbivorous lizard may have been in the habit of browsing, sustaining itself, kangaroo-like, in an erect position on its back extremities and tail.'[3] But *Hadrosaurus* was a reptile and not a mammal, a fact that clearly troubled Leidy, and he cast about for a reptilian model that would allow him to leave his hadrosaur in a prostrate pose. The solution was presented by another of Owen's Crystal Palace restorations, the *Labyrinthodon*. Hawkins had taken it upon himself to restore *Labyrinthodon* looking like a man-sized frog and Leidy seized upon this as the only viable alternative to the kangaroo in the cold-blooded domain. It is not improbable, he concluded, thus covering all possible options, 'that *Hadrosaurus* retained the ordinary prostrate condition, progressing in the manner which has been suspected to have been the case in the extinct batrachian [amphibian] of an earlier period, the *Labyrinthodon*'.[4] The Triassic amphibians, in fact, shared none of the peculiar specialisations for hopping that we observe in a frog. When the rest of its body appeared, *Labyrinthodon* was shown to be a crocodile-shaped beast and it probably lived a similar lifestyle in Triassic rivers and swamps. Leidy soon abandoned the frog and resorted to the kangaroo analogy.

Leidy's suspicion that the bulky hadrosaur anticipated kangaroo-style locomotion was fully supported when another saurian emerged from the New Jersey marl pits, although this was to fall trophy to one of his former students, Edward Drinker Cope. As a precocious child Cope's abiding interest had been reptiles; by the age of ten he was producing accurate scale drawings of the intricate skull bones of the ichthyosaurs in the Philadelphia Academy. Nevertheless, his conduct at school consistently fell far below that expected of a Quaker. His love of a brawl, both physical and intellectual, was to resurface with dramatic consequences in later years when he became embroiled in his infamous battle with the Yale palaeontologist O. C. Marsh. Cope was to prove the driving force behind the quest for Mesozoic saurians for three decades in the New World;

'driving' is the operative word, for his nature was one of urgency, drive and ambition, mixed with pride and jealousy: an explosive combination. He possessed a fiery temperament, a fighting spirit and an inordinate capacity for research. In 1858, at the age of 18, Cope issued his first challenge to his Philadelphia elders, submitting a paper to the Academy on the finer points of salamander classification. It was immediately recognised and published. In only three years an additional *thirty* papers appeared on snake, lizard and amphibian classification, at the end of which time he *began* his formal zoological education by attending Joseph Leidy's course in comparative anatomy at the University of Pennsylvania. (Cope published prolifically – 1,400 papers, books and monographs during his lifetime – turning out papers with such rapidity that numerous errors crept in, although pride prevented him acknowledging them.) Cope's ability quickly manifested itself to the elders of the Academy, and in 1861 he was elected a member, and four years later its Curator. With the onset of the Civil War, Cope, notwithstanding the Quaker pledge of pacifism, became deeply committed, causing his father to pack him off to Europe lest he should join forces with the Union. Returning from his European tour in 1864 he was appointed Professor of Zoology at Haverford College, while still only 24. Cope's highly strung disposition made him singularly unsuited to a sedentary college existence, neither would his uniquely individualistic nature permit the university to swallow him up; teaching had consumed so much of Leidy's time that little remained for research, and Cope had learned the lesson. Besides, Leidy's lectures on the extinct giant saurians like *Hadrosaurus* had fired Cope's imagination and he yearned to live in the Haddonfield district where the monsters lay interred. He made frequent trips to the dinosaur-bearing New Jersey marl pits and in 1866 bought a house in the area. So successful was his collecting that two years later he resigned his college post and, selling the farm that his father had presented him (in the vain hope that he would acquire a taste for farming), 'retired' to Haddonfield in order to devote himself fulltime to the quest for new saurians.

In the summer of 1866, on one of his periodical visits to the marl pits of West Jersey, he made his first dinosaur find, a hadrosaur thigh bone some thirty inches long. He deliberately made himself a familiar figure around the quarries to ensure that all fossil strikes were reported to him as soon as they were made. The policy worked. Later that same summer, when one of the pits produced a large quantity of saurian remains, Cope was hastily summoned. The bones had generated a flurry of excitement, and Cope wrote to his father of the sight that greeted him in the bottom of the pit:

> I found the remains of much greater interest than I had anticipated – being nothing more or less than a totally new gigantic carnivorous Dinosaurian probably of Buckland's genus *Megalosaurus*! which was the devourer and destroyer of Leidy's Hadrosaurus, and of all else it could lay claws on. One claw or ungual phalange is preserved and is 8 inches long at the least; it is in form between the talon of an eagle and the claw of a lion.[5]

This discovery was to establish Cope's reputation internationally. The Cretaceous *Hadrosaurus* from the New Jersey Greensand had at last met its match. 'The discovery of this animal filled the hiatus in the Cretaceous fauna,' wrote Cope, adopting the prevailing gladiatorial ethic, 'revealing the carnivorous enemy of the great herbivorous Hadrosaurus.'[6] The hadrosaur's adversary was not the English megalosaur, however, but a more formidable protagonist. It was, said Cope, taking a closer look at the dagger-like teeth and raptorial claws, the most rapacious terrestrial carnivore the world has known. Exhibiting the dinosaur before the Academy in August, 1866, he selected the name *Laelaps aquilunguis* for the beast of prey. (In Greek mythology, Laelaps was a fleet-footed dog given by the goddess Diana to the youth Cephalus, who loved to hunt. While giving chase one day Laelaps was turned to stone by the gods, its frozen pose being so lifelike that it appeared on the verge of leaping.)

Fortunately, Cope's skeleton included both arm and leg bones and the size discrepancy was even more accentuated than in Leidy's saurian. The thirty-one inch thigh bone dwarfed the corresponding upper arm bone, which might have stretched to twelve inches when restored (Cope's bone was incomplete). This bore out Leidy's belief that the animals were giant reptilian kangaroos. The Cretaceous scene was now set: both *Hadrosaurus* and *Laelaps* leaped like monstrous kangaroos, even though they were well over twenty feet long, and pursued each other in this fashion. The ridiculously small front legs in *Laelaps* were useless as weapons in the attack; that function was relegated to the eagle-clawed hind feet. The *Laelaps* leaped in pursuit of the hadrosaur whilst simultaneously wounding it with its talons! Or so Cope thought.

Cope's discoveries were made at an opportune moment. His resurrected dinosaurs were soon to be returned to their supposed life-like appearance as their contemporaries had been in London a few decades before. The fame of the Crystal Palace restorations had spread across the Atlantic closely followed by their inventor, Waterhouse Hawkins. The Board of Commissioners of Central Park in New York were acutely aware of the educational value of the London venture. Andrew Green, the able Board administrator, observed in 1868 that 'although the Sydenham Park is situated several miles from the metropolis, and notwithstanding that there is a charge for admission, yet hundreds of thousands of people have visited it annually for the last fifteen years, while the animal restorations have been a permanent element of attraction, and a source of valuable instruction to multitudes who would have gained this kind of knowledge in no other way'.[7] Like Hawkins, Green believed in education through 'object teaching', and he was the driving force behind the founding of the American Museum of Natural History in 1869. In May 1868 Green approached Hawkins in New York with a request 'to undertake the resuscitation of a group of animals of the former periods of the American continent'. Hawkins could not resist the offer and quickly set to work, visiting museums in Washington, New Brunswick, Albany and New Haven in search of suitably dramatic subjects. In Philadelphia he met Cope and Leidy at the Academy, and his quest ended upon seeing the giant saurians reposing in the Academy's Museum.[8] Leidy's insistence on the

8. *Megalosaurus*, the Jurassic carnivorous dinosaur, as envisaged by Owen in 1854 (left), contrasted with Cope's picture of the carnivore *Laelaps* (opposite) less than two decades later. Palaeontologists working in America were the first to appreciate that the flesh-eaters walked on their hind limbs.

tooth and claw struggle for survival in prehistoric times, a view amplified and made all the more bloodthirsty by Cope, was obviously a great influence upon Hawkins and it was embodied in the poses he planned for his creatures.

From September until December, 1868 he took moulds of the museum's fossil *Hadrosaurus* and *Laelaps* and filled the gaps left by missing bones with plaster replicas. While Hawkins was at the museum Cope's collectors in Kansas shipped back to Philadelphia a forty foot snake-necked plesiosaur found in the Chalk and it was apparently Hawkins himself who spent a month 'extricating it from the matrix' for Cope, before taking casts for his own proposed museum in Central Park.[9]

This long-necked plesiosaur, described by Cope in 1868 and named *Elasmosaurus* (or 'ribbon reptile'), sparked off the most explosive of all palaeontological feuds, a vendetta that was to shape American palaeontology for the next twenty years. Cope thrived on controversy but was a bitter loser. So his friendship with Othniel Charles Marsh, Professor of Palaeontology in the newly opened Peabody Museum of Natural History at Yale University, turned abruptly to antagonism and finally bitterness when Marsh dared criticise Cope's restoration of the elasmosaur.

Unlike his talented and prolific contemporary, Marsh waited thirty years before publishing a paper. His childhood had been backward and unadventurous, and even at the age of 21 he looked forward to a secure if mundane

9. With the kangaroo analogy his only viable option, Cope pictured his formidable *Laelaps* as a leaping dinosaur.

career in a mechanical trade; not an auspicious start for a future leader of American science. The catalyst that initiated the belated take-off was money. Marsh was nephew to millionaire George Peabody, whose successful London banking deals and philanthropic tendencies made him a natural target for needy relatives. Peabody, aware of his own lack of a sound education, donated the lion's share of his fortune to educational institutions. Peabody, concluded Marsh shrewdly, would never part with money to an *unaspiring* nephew. So, in an attempt to impress his uncle, Marsh – who had entered Phillip's Academy in Andover, a prep school, at the age of 21, where he assumed the nickname 'Daddy'! – began hoarding honours, awards and good reports. These he forwarded to his distinguished uncle, adding a cautious request for money to enable him to continue his education at university. Peabody responded to the initiative and sent an ageing Marsh (he was now 25) to Yale College.

Marsh passed the Civil War years in Europe, examining the fossil collections and calculating his next move. Cope, on his whistlestop tour of European capitals, exploiting the vaults of every museum and acquainting himself with everyone of importance, made a point of visiting Marsh at Berlin University in 1863; this was their first encounter. Marsh was 32, Cope nine years his junior; Marsh, the late starter, had barely begun to contribute, Cope had a multitude of scientific papers to his credit. Nothing tangible divided them, yet Marsh found Cope's 'eccentricities of conduct' hard to tolerate. In later years Marsh claimed that even then he doubted Cope's sanity – a recollection clouded by an intervening lifetime of mutual hatred. It was a classic clash of personalities: the brash, energetic Cope, imaginative and excitable, and quick to state an opinion, coming up against the stolid Marsh.

Marsh had in fact spent the years abroad hatching a scheme to obtain still more money from his uncle. He hoped to use his influence to persuade Peabody to found a museum of natural history at Yale, and thereby have himself installed as professor! The gambit worked. Marsh stunned his Yale teachers with the announcement that his uncle was to donate $150,000 for the founding of a museum and a chair of palaeontology. By way of gratitude, the authorities were only too pleased to appoint Marsh as professor. With an income for life guaranteed by Peabody, Marsh was free to research and explore. From this vantage point he moved from strength to strength, acquiring a taste for prestige only to be satiated by new appointments; and with his nomination as Vertebrate Palaeontologist to the United States Geological Survey, Marsh found his funds swelling as well. While Cope's dwindling finances in later years forced his retreat into the confines of a home doubling for museum and laboratory, Marsh enjoyed all the resources of the spacious Peabody Museum, access to the Survey (which sponsored his field parties and paid his Yale staff), and the luxury of an 18-room mansion in New Haven. While Cope's papers proliferated, Marsh's prestige rose, and both men eyed each other suspiciously. Marsh, whose scientific communications were always well planned and executed, laboured methodically and single-mindedly, with the result that his output was but a fraction of Cope's. Cope, ever the opportunist, flitted from subject to subject, unable to ignore any material that came his way; his papers dealt with everything from fossil sharks to his own theory of the evolutionary mechanism. Marsh, sadly no theoriser, stuck to solid descriptive work, disclosing his results in the dryest palaeontological prose. Marsh could indeed afford to despise error. Cope, often in haste to seize priority, blundered more than once. It was on this point that the two headstrong protagonists finally clashed openly.

Cope's published descriptions of the 'ribbon reptile' made it appear the most remarkable creature of all time, one so different that it required the setting up of a totally new order of reptiles – the Streptosauria, or 'twisted reptiles', so named after the long flexible neck – to contain it. Cope's sketch restoration of this 'wonderful creation', as Marsh sarcastically styled it, appeared in the Transactions of the American Philosophical Society. The sketch was based on the mounted elasmosaur skeleton in the Academy museum, a display that Cope proudly showed off to Marsh on the latter's visit to Philadelphia in 1870. Cope was unable to gloat over his prize possession for too long. When Marsh saw the skeleton and had its peculiarities explained to him by Cope, he immediately perceived the reason for the creature's uniqueness. 'I noticed that the articulations of the vertebrae were reversed,' Marsh confided later, 'and suggested to him gently that he had the whole thing wrong end foremost.'[10] Cope was furious and let Marsh know it in no uncertain terms. He had not studied the animal for years without coming to recognise one end from the other.

It seems he had, for Professor Leidy quietly took the last vertebra from the 'tail' and fitted the skull to it. Cope had indeed placed the head on the end of the tail to produce the wondrous creature of his restorations. When Marsh brought up Leidy's similar misgivings, Cope's 'wounded vanity received a shock from which he has never recovered, and he has since been my bitter enemy', said Marsh later. An embarrassed Cope quickly endeavoured to buy up, at his own expense, all those Transactions containing his restoration. Marsh, of course, held on to two copies. From this moment on Marsh and Cope were implacable foes, competing in every venture and fighting continually over publication priorities. For two decades American palaeontology was dominated by this feud and the motivation spurring on both men in their quest for fossils must be seen as a product of this enmity. The war did not really get under way until 1877. In that year Marsh announced to the world his discovery of a really giant dinosaur, the 'titanic retile' or *Titanosaurus*. In the same paper, which appeared in July, he took the opportunity to point out that Cope's special beast *Laelaps* bore the same name as at least two other creatures. Since these had been named at least thirty years earlier, they had priority. According to the rules governing name-giving, '*Laelaps*' would have to be changed, and Marsh himself renamed it *Dryptosaurus*.[11] Cope paid no attention to Marsh's correction and continued to use *Laelaps* until his death. But within days of Marsh's paper Cope had retaliated by claiming that *Titanosaurus* itself was a preoccupied name, and that Marsh had better find a new one![12] In this hysterical vein the 'debate' continued while other palaeontologists, trying not to be drawn into the fray, steered a more sober course.

10. Hawkins, seen standing below the *Hadrosaurus* skeleton in the Central Park studio, had made good progress in the early stages of reconstruction. This illustration, sketched in 1869, confirms that he had completed one sphinx-like *Hadrosaurus* (left) as well as a pair of giant elk.

Meanwhile, the original culprit, *Elasmosaurus*, as well as the other Academy saurians, had been drawn and cast by the indefatigable Hawkins. The Philadelphia Academy provided him with facilities and equipment and in exchange they received the first full-size skeletal restoration of *Hadrosaurus*, thirty-nine feet long and standing on its hind feet with front legs grasping a tree as it browsed off its leaves. This model remained on exhibition at the Academy for many years before being broken up. In December, 1868 Hawkins returned to New York and commenced the work of restoration in a palaeontological studio specially erected in the south-west corner of Central Park. There he began to build two models of *Hadrosaurus*, one standing like the Academy's specimen and another reclining in a sphinx-like pose, with oversize hind limbs and bird-like feet.

The giant reptiles were to be housed in the Paleozoic Museum planned for the park. Like its London counterpart, the building was designed as a huge iron-framed glass building with an arched roof. Long before work on the museum had begun Hawkins had sketched the interior, and from this we can glean some idea of the poses he had in mind. The larger *Hadrosaurus* in the sketch was being set upon by one of the smaller but ferocious *Laelaps*, while two other *Laelaps* squabbled over an enormous corpse. Giant plesiosaurs and mosasaurs lurked in the water at their feet. Models of early man's more spectacular contemporaries, such as the mammoth, giant elk, giant armadillo and giant sloth were also planned. The aim, according to the commissioners' *Report* that year, was no less than 'a complete visual history of the American continent from the dawn of

11. The Paleozoic Museum in Central Park, as it would have looked upon completion. On the left one *Laelaps* attacks a *Hadrosaurus*, while two other *Laelaps* squabble over a corpse. In the foreground skulks the snake-necked *Elasmosaurus*. Giant armadillos and ground sloths are visible on the right. The Museum was to have been a grand iron-framed, glass-curtain building, much like the Sydenham Crystal Palace.

creation to the present time'. Although falling far short of this, the museum would nevertheless have graphically illustrated those periods of earth history when giants truly reigned.

Foundations were laid by America's foremost landscape gardener, Frederick Law Olmstead, opposite Sixty-Third Street, and maps of the park began appearing prematurely with the museum building already depicted.[13] The events that followed are difficult to piece together because of the highly fragmentary nature of the evidence reaching us. For several years William Marcy Tweed had been building up his private empire in New York. The Tweed Ring had bought or bullied its way into complete control of all the bastions of power in the city, from the Democrat Party headquarters to the courts. Tweed henchmen were manoeuvred into all key executive posts. 'From 1866 to 1871,' the historian A. B. Callow has recently said, the Tweed organisation 'plundered the City of New York with such precision that it has received the singular distinction of being labelled the model of civic corruption in American Municipal history.'[14] Graft and treasury plundering left 'Boss' Tweed immensely rich, which in turn increased his stranglehold over the city, and he became the third largest owner of real estate in New York. Tweed retained his political power through patronage, providing jobs in return for continued support. He needed to control large sections of the city to accommodate these supporters and it is hardly to be wondered at that he soon cast an envious eye in Central Park's direction. In 1870 he bulldozed a piece of legislature through to abolish Andrew Green's commission and placed his own men in control. Green stood helpless as Tweed's right hand man Peter 'Brains' Sweeny took control of the park. The *New York Times*, one of the few institutions able to stand up to the Ring, prophesied the systematic exploitation and ensuing demise of the park.

The new Tweed administration, which took office on May 1, 1870, was hostile to the Paleozoic Museum from the outset. Ostensibly, work was halted because it was costing the taxpayer too much money. The estimated final cost of the building was $300,000, which was 'deemed too great a sum to expend upon a building devoted wholly to palaeontology – a science which, however interesting, is yet so imperfectly known as not to justify so great a public expense for illustrating it'. It is ironic in the light of this pronouncement that the Ring parted with $8 million in the first 18 months as patronage to keep *four thousand* labourers in the Park, where before Green had needed only $¼ million a year. Some months later, a speaker at the Lyceum of Natural History in the American Museum, one of an ever increasing number of Tweed adversaries, satirised Tweed's position: 'Professor Hawkins,' he said imitating Tweed, 'has been studying books and bones; what does he know about the management of Central Park; he was only an Englishman anyway, and the idea of his trying to get a museum in this City without a corresponding scheme for dividing the profits, was an absurdity.'[15] Since the Ring foresaw no financial gain, it was decided to fill in the foundations.

By early 1871 Hawkins' models were nearing completion in his studio. An illustration in the old commissioners' *Report* shows that even in 1868 he had completed the smaller *Hadrosaurus* and cast the skeleton of the larger one. He had also probably finished the snake-necked elasmosaur. An inventory of park property made by the Ring the day following the takeover shows that he had built the rapacious *Laelaps* and 'seven plaster casts' of unspecified type. But they were not to survive. One day in the spring of 1871, vandals, under orders from the Ring commissioners and with the blessings of Sweeny, broke into the studio and smashed the dinosaurs with sledge hammers, taking away the pieces to bury in the park. The moulds and sketch models were likewise destroyed in an act of sheer wanton vandalism, while the shocked Hawkins was taunted that 'he should not bother himself about dead animals, when there were so many living ones to care for'.[16] The motivation for this destruction is lacking. 'Boss' Tweed could have lined his own pocket by selling the models to the Smithsonian Institution in Washington, which would gladly have taken them. Tweed's contempt for the project may have stemmed from a deep-rooted religious prejudice. Certainly he appeared sceptical of the great antiquity of dinosaurs themselves, referring to Hawkins' models as 'specimens of animals *alleged* to be of the pre-Adamite period'.[17] Tweed's downfall followed a few months later at the hands of Louis Jenning, an English editor hired specifically for the purpose by the *New York Times*. The *Times'* invectives and exposure of the frauds led to the Ring's loss of political power and the prosecution of its leaders. Tweed and Sweeny fled abroad and the Park commission reverted back to the capable hands of Andrew Green. But by then the damage had been done.

40

A shattered Hawkins retreated to the College of New Jersey (Princeton University) and academic seclusion for a few years before returning to his native England in 1877. Tweed's animosity towards Hawkins' godless creations could have had no lasting influence, for in 1875 Hawkins began work again and in the space of only two years had finished (or nearly finished) a series of seventeen large paintings for the College. The prehistoric scenes, which owe much to his collective Crystal Palace and Central Park researches, are astonishing for his attempt to infuse some life into the resuscitated dinosaurs.[18] He even made one final effort to reanimate the fated *Hadrosaurus*, interrupting his paintings in 1878 to construct a lifesize effigy for the centenary of the Declaration of Independence celebrated by an International Exposition at Philadelphia. Since the model was commissioned by the National Museum in Washington it was later returned to the capital and set up in the open air in front of the museum. But not even this model was destined to survive. It seems that it had been made of plaster and as a consequence of being exposed to the elements it slowly weathered away until it was finally destroyed.

One outcome of the discovery of kangaroo-like dinosaurs was that many looked disparagingly at the Crystal Palace models, looking as they did like reptilian rhinoceroses. Marsh was a particularly harsh critic. He was unable to see the restorations in an historical light – as a product of the limited knowledge available to Owen and Hawkins. After visiting the Palace in 1895 he made some deprecating remarks concerning the models at a British Association meeting:

> The Dinosaurs seem . . . to have suffered much from both their enemies and their friends. Many of them were destroyed and dismembered long ago by their natural enemies, but, more recently, their friends have done them further injustice in putting together their scattered remains, and restoring them to supposed lifelike forms. . . . So far as I can judge, there is nothing like unto them in the heavens, or on the earth, or in the waters under the earth. We now know from good evidence that both *Megalosaurus* and *Iguanodon* were bipedal, and to represent them as creeping, except in their extreme youth, would be almost as incongruous as to do this by the genus *Homo*.[19]

The monsters may have appeared as curious relics in Marsh's time but his criticisms were unjust: there was no reason to suspect in 1854 that many dinosaurs looked like gigantic kangaroos.

In fact, English palaeontologists had quickly responded to the American revelations and a full acceptance was not long in the coming. Thomas Henry Huxley, lecturer extraordinary in palaeontology at the Royal School of Mines in Jermyn Street (later to become London University's Imperial College), seized upon Leidy's findings and proposed that English donosaurs such as *Iguanodon* must also have supported themselves on their hind limbs. More than this, three-toed footprints found in the English Weald (made by a large creature with a long stride: presumably *Iguanodon* itself) led him to the assumption that this creature *walked* upright on its rear legs. Huxley, though, as an ardent evolutionist, had ulterior motives for accepting Leidy's beliefs. No sooner had the *Origin of Species* been published in 1859 than Huxley's tenacious defence of its tenets against palaeontologists and pulpit alike earned him the title of 'Darwin's Bulldog'. After Darwin had prepared the groundwork and seen the *Origin* through the press he retired from the arena and left Huxley as his champion. Anything therefore that would further the Darwinian cause was grist for Huxley's mill. Where Darwin had feared to tread, Huxley rushed in, eager for battle. Huxley had no qualms about removing man from his entrenched position in the Universe. Man's evolutionary origin was an inflammatory subject, guaranteed to raise an indignant outcry from the pulpit, and Huxley delighted in baiting his opposition. The bewhiskered, stern Victorian scowling for his portrait hardly tells of the engaging *provocateur* lurking inside. No one was spared Huxley's taunting; prime ministers, scientists and clerics, all were dispensed with equal relish. 'Extinguished theologians,' he jibed in 1860, 'lie about the cradle of every science as the strangled snakes beside that of Hercules.' He devoted increasing time in later years to polemical works, carrying the evolutionary banner into his opponents' camps with such success that he conferred upon himself the title of Episcophagous or 'bishop-eating'. He was the first to fully grasp the implications of evolution for man – and he was the first agnostic. Deftly dispensing with clerical opposition, however, was only an outlet for Huxley's mischevious leanings; it was in the battle against his academic elders, and particularly in his use of fossils as evidence for evolution, that Huxley served the evolutionary cause best.

So it was that in one of his extremely popular Royal Institution lectures, Huxley in 1868 began his onslaught on the major palaeontological problems confronting the doctrine of evolution. How was it, he argued on behalf of his protagonists, that if all the animals proceeded from a common stock great gaps existed between the classes, such as between fish and reptiles, reptiles and mammals, and so on. As an evolutionist, he explained, he believed that there were once linking forms alive but that they had since died out. For example, there seems to be a great gap between reptiles and birds, yet the then newly discovered *Archaeopteryx* or 'ancient feather' from Solnhofen in Bavaria was a perfect half-reptile, half-bird that lived in Jurassic times: it bore feathers, yet it had teeth, a long bony tail and long fingers. Birds hop or run on their hind limbs and so Huxley cast about for a reptile that did the same and that could meet the birds half way. The *Iguanodon* with its three-toed hind foot was just the creature. The erect *Iguanodon* and its allies were endowed with hindquarters that 'wonderfully approached' those of birds, supplying Huxley with his first clue to the ancestry of the birds. The little *Compsognathus* dinosaur, a contemporary of the hybrid *Archaeopteryx* in the Bavarian slate deposits, was still closer to the bird in appearance. Since it was only two feet long, it also approached the early birds more closely in size and thus rendered counter-arguments bearing on *Iguanodon*'s gigantic size less objectionable. Long, delicate hind limbs gave *Compsognathus* a quite avian appearance. 'It is impossible to look at the conformation of this strange reptile,' said Huxley during his lecture to the Royal

Institution, 'and to doubt that it hopped or walked, in an erect or semi-erect position, after the manner of a bird, to which its long neck, slight head, and small anterior limbs must have given it an extraordinary resemblance.'[20]

Huxley's absolute commitment to evolution led to his quick appreciation of Leidy's discoveries and thus hastened their acceptance in England. Huxley was then in a position to reinterpret Edward Hitchcock's observations of 'bird' tracks in the ancient Triassic New Red Sandstone of Connecticut and Massachusetts. Hitchcock was President of Amherst College, as well as its Professor of Natural Theology and Geology, and it was through his intervention that the college first acquired some extraordinary three-toed fossil tracks in 1835. These he deposited in his 'cabinet', as the collection of fossils was known. The tracks had been spotted in the valley cut by the Connecticut River and had been taken to Hitchcock for identification. Realising their worth, he laboured for three decades, wandering the eighty-mile stretch of the valley until he had uncovered twenty-one footprint sites. By 1848 he had characteristic tracks of forty-nine species, of which thirty-two were erect, bipedal creatures with only three or four toes. Since the tracks so resembled those of wading birds, Hitchcock named them Ornithichites or 'stony bird tracks'. Upon the suggestion of James Dana, later to become Yale University's Darwinian stalwart, Hitchock allotted names to all his tracks to avoid confusion, dividing them according to such characteristics as number and thickness of toes, presence of heel, and so on. Some of his colleagues objected that it was not sound practice to name creatures where no actual relics had come to light. No skeletons were ever found but Hitchcock argued persuasively that particular feet characterised particular animals: there would be no confusing ostriches and eagles or carnivores and ruminants if only their tracks were known. Indeed, Cuvier's comparative anatomy had made such an approach respectable. From a single bone the French Baron had rebuilt an entire animal and likewise from the footprint alone he could quickly conjure up its maker. 'Any one,' said Cuvier, 'who observes merely the print of a cloven hoof, may conclude that it has been left by a ruminant animal, and regard the conclusions as equally certain with any other in physics or morals.' From a footprint Cuvier's comparative technique permitted him to vividly recall the creature even to the point of describing its teeth. With this impressive precedent, Hitchcock set to work, describing the creatures that in Triassic times had swarmed through the Connecticut valley leaving only footprints for mementoes. Many of the four-footed trails were supposed left after the *Labyrinthodon*'s passing, but in the main the prints were three-toed and very large. Hitchcock's 'giant animal', *Brontozoum giganteum*, known only from its tracks, was by far the dominant creature on the landscape, in terms of both numbers and size: its foot reached eighteen inches in length and the stride was anything up to six feet. These giants, said Hitchcock in 1848, were obviously wading birds four or five times the size of ostriches. 'They must have been giant rulers of that valley,' thought Hitchcock, a conclusion apparently confirmed by their number, for he believed that many flocks had lived in the region. 'Their gregarious character appears from the fact, that, at some localities, we find parallel rows of tracks a few feet distant from one

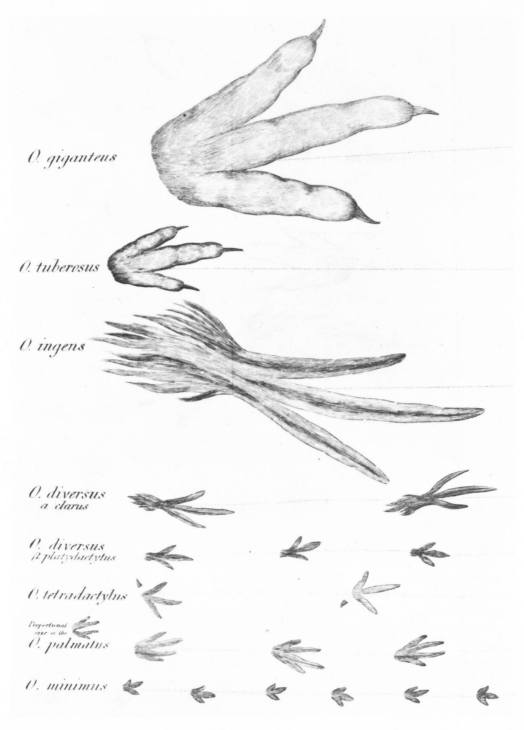

O. giganteus

O. tuberosus

O. ingens

O. diversus
a clarus

O. diversus
β platydactylus

O. tetradactylus

Proportional
size of the
O. palmatus

O. minimus

12. Hitchcock's 'stony bird tracks' from the Connecticut Valley. The largest was over 18 inches long. Hitchcock died in 1864 believing that the prints were made by Triassic birds.

another.' Like many wading birds, flocks of these creatures had strutted majestically along the shoreline of the ancient sea.

> Whatever doubts we may entertain as to the exact place on the zoölogical scale which these animals occupied, one feels sure that many of them were peculiar and gigantic; and I have experienced all the excitement of romance, as I have gone back into those immensely remote ages, and watched those shores along which these enormous and heteroclitic beings walked. Now I have seen, in scientific vision, an apterous bird, some twelve or fifteen feet high, – nay, large flocks of them, – walking over the muddy surface, followed by many others of analogous character, but of smaller size. Next comes a biped animal, a bird, perhaps, with a foot and heel nearly two feet long. Then a host of lesser bipeds, formed on the same general type; and among them several quadrupeds with disproportioned feet, yet many of them stilted high, while others are crawling along the surface, with sprawling limbs. . . . But the greatest wonder comes in the shape of a biped batrachian, with feet 20 inches long. We have heard of the Labyrinthodon of Europe, – a frog as large as an ox; but his feet were only 6 or 8 inches long, – a mere pygmy compared with the *Otozoum* of New England. Behind him there trips along, on unequal feet, a group of small lizards and *Salamandridae*, with trifid or quadrifid feet. Beyond, half seen amid the darkness, there move along animals so strange that they can hardly be brought within the types of existing organization. Strange, indeed, is this menagerie of remote sandstone days.[21]

Hitchcock's prehistoric zoo was modelled after existing types and his first and last impression was understandably that the Triassic was ruled by monstrously tall ostrich-waders. 'But,' he had once said unintentionally punning, 'the geologist should be the last of all men to trust to first impressions.'[22] Yet Hitchcock himself clung obstinately to his Triassic ostriches even in face of Leidy's unearthing the kangaroo-mimic *Hadrosaurus*. In England, the younger, more adventurous Huxley had no such qualms about repopulating the Connecticut valley.

Huxley was in a much stronger position. The wonderful evidence of life in Triassic times afforded by these sandstones, notwithstanding the lack of bones or feathers, was full of instructive suggestion about the creatures that had once traversed these sandy beaches. 'The important truth which these tracks reveal,' affirmed Huxley, 'is, that at the commencement of the Mesozoic epoch, bipedal animals existed which had the feet of birds, and walked in the same erect or semi-erect fashion.' The smaller tracks Huxley supposed *were* left by birds, but many of the remainder (especially the giant three-toed prints) were those of dinosaurs. Thus Huxley was willing to admit that large bipedal dinosaurs were alive *even in distant Triassic days*.

In fact, all the tracks are attributable to archaic reptiles and amphibians, birds had yet to make an appearance in the skies, although Huxley was not to know this any more than Hitchcock. Nevertheless, Huxley's point was well taken: the confusion between the tracks of birds and bipedal dinosaurs served only to highlight the *similarities* between these major groups. If by tracing birds back,

Huxley could show that they became so reptilian as to be indistinguishable from the more avian dinosaurs, then he had demonstrated that there was 'nothing wild or illegitimate' in the hypothesis that birds had their ancestral roots within the dinosaurs. This was probably Huxley's first major achievement in bringing palaeontology to the aid of evolution; an accomplishment that was quickly surpassed by Huxley's enumeration in 1870 of the steps in the evolution of the one-toed horse from its archaic four or five-toed ancestor living in the Eocene.

Huxley's faith in *Iguanodon*'s bipedality was totally vindicated in 1878 when an entire troupe of iguanodonts emerged from a coal mine in Belgium. The dinosaurs were interred over a thousand feet underground at Bernissart, near Mons, not in the coal itself but in a fissure filled with Cretaceous marls. The *Iguanodon*'s Wealden world was already 180 million years removed from the coal-forming swamps of the archaic Carboniferous period. However in Cretaceous times a deep ravine in the coal deposits had acted as a lethal trap to the dinosaurs. Into this ravine had fallen, in quick succession, no less than thirty-one fully grown *Iguanodons*. The miners cutting a new shaft in the Fosse Sainte-Barbe mine had tunnelled out of the coal and into this choked Cretaceous ravine, completely destroying one skeleton before they realised. The enormity of the find only became apparent when experts from the Brussels Royal Museum of Natural History tried to drive a new shaft 100 feet below the original one to extract the fossils from below, only to come up against a vertical continuation of the cemetery. The ravine was literally filled for hundreds of feet with stranded *Iguanodons*, all of which had been unable to escape and were entombed in mud as the fissure filled up with flood water. With the blessings of an enlightened mine Director, the shafts were requisitioned for three years as laboratory staff and assistants moved underground to hew out blocks of marl-encapsulated bone for dispatch to the museum. Here Louis Dollo, still only in his mid-twenties, was given the task of preparation and description, to which end he devoted the rest of his life. By 1883 he had mounted his first skeleton and by the turn of the century four more erect skeletons had been added; imposing creatures standing on their rear legs and with front legs free to grasp trees. Currently, the Royal Museum houses thirty-one whole and partial skeletons extracted from the mine, forming one of the most sensational group displays anywhere in the world. This incredible recreated menagerie mustered in one spot allowed great scope for a comparative study of the variations within a *single* species of dinosaur. Dollo actually thought he had two species, distinguished from one another by size, but these are more probably sexual differences. Strangely, the individuals that fell into the ravine were all fully grown adults, no youngsters perished with their parents. Dollo's fertile imagination conjured up a picture of an *Iguanodons*' graveyard where the decrepit and senile animals would slink off to die, much like our mythical elephants' graveyard.[23]

Iguanodons were undeniably some of the most successful dinosaurs in the group's history. All were large, but some attained really gigantic proportions. The American Museum Sinclair expedition to the Mesaverde coal measures of Wyoming and Colorado in 1937 returned armed with colossal iguanodont tracks

13. *Iguanodon* has become one of the best known dinosaurs, largely as a result of Louis Dollo's studies of the Bernissart specimens.

hewn out of the coal. The prints, each just under three feet long, occurred at 15 feet intervals, indicating a creature with a huge stride. On the basis of these figures the mysterious erect beast was thought to tower to a height of 35 feet. But the dinosaur itself remains an enigma: its skeleton was never found.[24]

Not all the erect plant-eaters reached giant proportions. One of the most intriguing British dinosaurs is also one of the smallest. When fragments of it were first found in the Wealden Beds of the Isle of Wight, Mantell reasoned that the creature must have been a very small *Iguanodon* juvenile. Huxley, working with better material, recognised it as a miniature dinosaur adult in its own right and named it *Hypsilophodon*. This beast was only four or five feet long and was thus dwarfed by its *Iguanodon* contemporaries. Large eyes and an apparently opposable first toe in the foot led Othenio Abel, writing in Vienna in 1912, to suppose that this nimble plant-eater hopped from branch to branch, perching to eat fruit or leaves which it grasped with its small forepaws. Because *Hypsilophodon* was a primitive dinosaur, retaining a five-fingered hand and a foot with four very long grasping toes, Abel speculated that dinosaurian origins would be found among earlier bipedal tree-climbing reptiles. This roosting dinosaur naturally interested Gerhard Heilmann, whose speciality was bird origins, and he was at first inclined to agree with Abel about *Hypsilophodon*'s arboreal habits. He showed, however, that the first toe was not long enough to oppose the rest, so the dinosaur could not have grasped branches like a bird. The ancestry of the dinosaurs, he thought, would be better sought among agile ground-living bipedal reptiles. At first he likened *Hypsilophodon*'s foot to a monkey's and the dinosaur itself to a tree kangaroo, but later changed his mind and brought *Hypsilophodon* out of the trees altogether, thinking it better suited to life on the ground like its ancestors. But the tree kangaroo analogy had taken root and in 1936 W. E. Swinton reinstated *Hypsilophodon* to its former tree-climbing existence. Since it was not fleet of foot, he argued, this little dinosaur had taken to the trees as a refuge from the carnivores. *Hypsilophodon* is now quite well known from twenty whole or partial skeletons, all of which were found on the Isle of Wight. A thorough study of all the skeletons led Peter M. Galton of the University of Bridgeport in 1974 to allow the dinosaur to venture out of the trees once again.[25] Heilmann was right in thinking the big toe could not be opposed to the rest for perching, but Swinton was incorrect in his belief that *Hypsilophodon* could not run. Since it lived on the ground it had to be able to outpace carnivores and Galton's study of the limbs established, as will later be shown, not only that it could run, but that it was very fast indeed.

There was a spectacular flowering of large two-footed foliage browsers in late Cretaceous times, of which Joseph Leidy's *Hadrosaurus* was just one example of many. Indeed, *Hadrosaurus* and *Trachodon* were situated at the more normal end of the hadrosaur spectrum, for the group literally exploded into a diverse array of forms a short time after evolving from the iguanodonts. Hadrosaurs were an extremely successful group, flourishing in great numbers in North America towards the end of the dinosaurs' long reign. Consequently, hadrosaurs are some of the best known dinosaurs; skeletons are often unearthed completely articulated and more than one 'mummy', bound in shrunken skin tautly stretched across the corpse, has come to light. One extraordinary 'mummy' was found by Charles Sternberg in Kansas in 1908 and is now one of the most highly prized exhibits in the American Museum of Natural History. This 70 million-year-old corpse was found lying on its back with both legs kicked into the air as if in the throes of death. To preserve the beast with skin and organs intact the carcass

14. The mummified remains of *Trachodon*, housed in the American Museum.

48

must have been subjected to a particular sequence of events before fossilisation. The animal had obviously not succumbed to a predator since its body was intact; it must have died naturally. The dead animal lay on its back exposed to the sun for a lengthy period of time while dehydration occurred. As the carcass desiccated the hide and muscles shrunk around the body like the withered skin around a pharonic corpse, highlighting the rib cage and leg bones. The dehydrated 'mummy' was then quickly buried, perhaps carried downstream by floodwater and covered in fine sands and sediments before the skin had time to soften. The sands and clays thus moulded the brittle skin and duplicated all its prominent features. What we now possess is not the skin itself but an impression in the rock, albeit in perfect detail. Studying the 'mummy' we know that the hadrosaur was not covered with overlapping reptilian scales or bony plates, but by a mosaic of horny tubercles embedded in leathery skin. The tubercles differed in size over the surface of the dinosaur, with larger ones congregating on the sun-exposed back and legs and smaller ones on the belly.[26] These size differences may reflect contrasting colour patterns. There is no reason why dinosaurs should not have developed camouflage patterns like present-day mammals.

The herbivorous hadrosaurs were undoubtedly foliage croppers, to which end they developed a broad expanded duck-like beak that in life was covered by a horny sheath. The presence of this duck-bill long supported a belief that hadrosaurs were amphibious and that, anticipating ducks, they swam or waddled along in swampy regions, gobbling up soft aquatic vegetation at the water's edge. It was thought that in life a web of skin stretched between the fingers of the hand to facilitate swimming: these webs were actually 'seen' in the 'mummies', although such observations have always been hotly disputed.

The surprise occasioned by the sight of a beaked dinosaur was eclipsed by the discovery of the extraordinary prominences on the heads of many of them. When Charles Sternberg returned from the Red Deer region of Alberta in 1913, he presented the scientific world with its first crested hadrosaur. It was given to Lambe at the Canadian Geological Survey for study and he named it *Stephanosaurus*, although this was later changed to *Lambeosaurus*. This beast's skull bore an ornate crest rising from the crown of its head like a top hat.[27] As it turned out, hadrosaurs adorned in this fashion were by no means rare. Soon all manner of hooded creatures emerged from obscurity and it became quickly apparent that in late Cretaceous times in North America the group had undergone a period of accelerated evolution. Many diverse ornaments appeared: some like *Kritosaurus* were flat headed; others such as *Saurolophus* had solid domes; still more had bulbous 'Corinthian helmets' like *Corythosaurus*; and finally there were crests swept back like the plumes of a cavalier's hat, as in *Parasaurolophus*. All these ornate types appeared towards the end of the Cretaceous and many of the exotic crested duck-bills were discovered in Alberta by Canadian Geological Survey and American Museum expeditions. In 1920 Lambe was able to show that many of the ornate hoods were hollow and that the nasal passage followed a tortuous path through the crest, often following its bizarre contour closely – rising to the top of the helmet or following the plume

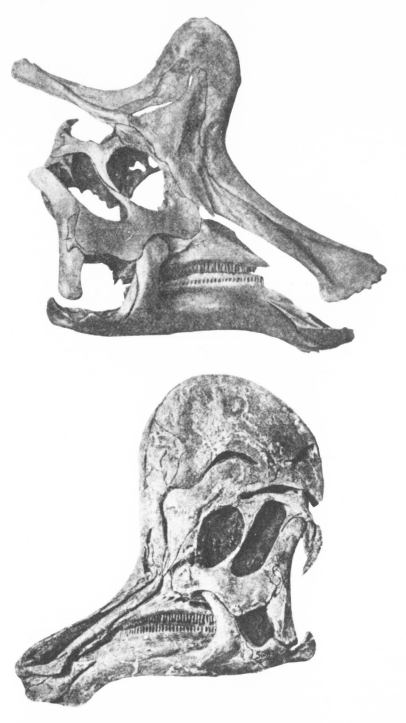

15. By 1920, Lawrence Lambe and Barnum Brown had unearthed an array of bizarre crested and helmeted hadrosaurs.

fully to its posterior extremity – before returning to the throat.[28] Why were some hadrosaurs flat-headed like *Hadrosaurus* while others were crested?

This question intrigued Baron Nopcsa – one of the most outlandish characters to have taken up the dinosaur's cause. Despite the 99 palaeontological publications to his credit, many of which were concerned with dinosaurs, Nopcsa is best remembered for his overriding obsession with Albania, its people, customs and geography, and it was his bizarre Albanian exploits that earned him a place in palaeontological legend. Indeed, as one obituarist observed, Nopcsa was 'Almost King of Albania'. Nopcsa, like so many brilliant people, was an extraordinary complex character whose fragile mental state seemed on occasions to have bordered on insanity. Before the First World War he lived the life of a baronial lord in Hungary, although his interests later focused on the Turkish province of Albania and he soon became an expert in that country's culture. At the cessation of the Balkan War in 1913, Albania became a pawn in the game played by all imperial powers with one another's satellites after military disengagement. The question of who should own Albania was hotly disputed. The Austro-Hungarians favoured the establishment of an independent state to satisfy nationalist demands, but sought to ensure that it was favourably disposed toward the Austro-Hungarian Empire by imposing a puppet king of their own choosing. Nopcsa, possessing the obligatory feudal background and armed with an overriding passion for the country, considered himself the most suitably qualified candidate for the post of monarch. He wrote to the army chief of staff requesting five hundred soldiers, a few cannons and two fast steamships. After establishing a beachhead he would quickly take the country and set himself up as King Nopcsa. After this he would simply marry the daughter of an American billionaire to raise the investment capital needed to refloat the economy! Instead, the Great Powers chose Prinz Wilhelm zu Wied as their puppet ruler, and the Albanians tolerated him for just six months before forcing him to flee for his life. During the First World War Nopcsa served in the Imperial Austro-Hungarian Army, not in any ordinary capacity, of course, but as a spy. Dressing as a Romanian peasant and growing his hair long, he carried out a series of daring exploits along the Romano-Hungarian border, while simultaneously writing paper after paper on dinosaurs and fossil reptiles. After the defeat, Nopcsa lost most of his estates without indemnification, so with declining finances he resorted to the life of a professional palaeontologist at the Hungarian Geological Survey. But the nobleman's arrogance made conflict with his more humble colleagues inevitable, and Nopcsa was only able to hold down the position for three years. With his secretary he set off on a motor-bike trip around Europe, covering many thousands of miles before running out of money. In 1933 he laced his secretary's tea with sleeping powder and then shot him, finally putting a gun to his own mouth and shooting himself.[29]

Interspersed in this comi-tragic life, as if on another almost esoteric plane, came an outpouring of articles, papers, even books, dealing with the reptilian inhabitants of the Mesozoic earth. Strangely, Nopcsa's scientific writings were very much down to earth and often pedantic in style. Nopcsa, the palaeotological renegade, the brilliant, if arrogant, misfit, was basically an ideas man. Thus in 1929, only four years before his suicide and already fearing another mental collapse, the Transylvanian baron pondered the problem of Lambe's odd helmeted and plumed duck-billed dinosaurs. Othenio Abel in Vienna had already suggested that these cranial ornaments were horns: Nopcsa's solution was a novel extension of this idea. Lambe had divided all his types into new genera and species, but Nopcsa retorted that all Lambe had done was to separate the males from the females. Hadrosaurs, he thought, exhibited elaborate sexual differences, manifested in the males by decorative bony ornaments. Thus the plumed 'species' *Parasaurolophus* was merely the male expression of the flat-headed *Kritosaurus*, *Corythosaurus* was the male *Diclonius*, and so on. Nopcsa was ready to eliminate half the known hadrosaurian species and would have done so if his ideas had gained any measure of popularity. He soon extended his revolutionary ideas to the flesh-eaters and iguanodonts. It did not worry him one bit that in Bernissart he seemed to have (using *his* criterion) one male *Iguanodon* amid twenty-three females. That was merely a reflection of the herding tendencies and highly polygamous nature of bipedal plant-eaters![30] To render the whole theory still more implausible it might be noted that only 'female' hadrosaurs have ever been found in some regions. In fact, these were separate species, there can be little doubt about that. Nopcsa's belief that the crest was totally ornamental ignored the most telling feature of it: that it was hollow and housed an elongated nasal passage. In the search for the function of the crest more would be gained by speculating on the possible benefits of such an elaborate nostril.

Since hadrosaurs were regarded as swimming and diving creatures, it was imagined that the nasal opening was placed on top of the crest to act as a snorkel allowing the animal to remain submerged. In Germany in 1938 Martin Wilfarth even suggested that the beak and hood together were only muscle attachment areas and that the whole structure had supported an elephant-like trunk. This device enabled the animal to snorkel air while feeding on lake bottoms, although it could double as a food-gathering device, as it does in an elephant. Wilfarth's bizarre dinosaur did not survive long. As Sternberg emphasised, developing a prehensile proboscis would completely undo any benefit derived from a horny duck-bill. Furthermore, there were no scars on the fossil skull to suggest muscles in life. Anyway, why develop a tortuous nasal passage *inside* the skull when there was a trunk for snorkelling outside?[31] In fact, there was no nasal opening on top of the skull; it was in the usual position on the tip of the snout. Wilfarth himself noticed this but escaped from its obvious conclusion by claiming that the snout opening was vestigial, an ancestral holdover covered by skin in life, whereas the real opening was on top. It is indeed easy to avoid one's beautiful theory being slain by an ugly fact. Sternberg himself believed that the S-shaped loop made by the nasal passage as it passed through the crest prevented water from entering the throat by acting like an air lock. This too is untenable for the simple reason that the water pressure around the immersed hadrosaur would have greatly exceeded the air pressure inside the crest and water would have forced its way in. Besides which, hadrosaurs undoubtedly used voluntary muscles to close the nostrils, an adaptation found in all aquatic creatures to prevent drowning.

For an aquatic, air-breathing dinosaur an auxiliary air supply would be a great advantage, allowing the animal to remain on the lake bottom for prolonged periods. So it is of no surprise that in the 1940s a school of thought developed around the idea that the helmets and bony plumes were aqualungs carried on the top of the head. Yet, when we come to measure the volume of this subsidiary air supply, it appears as a paltry 4% of the lung capacity in *Corythosaurus*. Furthermore, there would have been great difficulty transferring this vital air supply to the lungs. Drawing it into the lungs would create a vacuum in the helmet which could only have been relieved by the entrance of water if the animal was submerged, or air if the animal surfaced, both of which are self-defeating. In the first case the dinosaur would drown, in the second surfacing has rendered the aqualung superfluous.

The total inadequacy of all these explanations led John Ostrom, Professor of Geology in the Peabody Museum of Yale University, to look at the problem anew. Investigating the nasal apparatus and its function in modern reptiles, he concluded that a heightened sense of smell had been developed in some species. But it is never as acute as in mammals, where there are elaborate scrolls of bone coated with sensory skin to multiply the sensitive surface areas many times. Hadrosaurs, seemingly unable to neatly fold the skin around such scrolls of bone, had achieved the same end by merely lengthening the nasal passages and expanding the passages into chambers at points along their length. Both hadrosaurs and mammals had developed an acute sense of smell by increasing the total area of sensitive epithelium. The resulting structure in a mammal was compact and easily situated in the nose. In hadrosaurs the passages had to wind round the head and be protected by bony crests.[32] Why had hadrosaurs developed an acute sense of smell? Ostrom supposed that these plant-eaters were passive creatures, and since they were totally unarmoured they lacked all defence against powerful predators. Rather, they gained advance warning of approaching danger by rearing up on to their hind legs and sniffing for scents carried on the wind. They could thus move off before the flesh-eater had even come into view. A good sense of smell may have been necessitated by a loss of the characteristic reptilian method of predator detection. A lizard lays its jaws on the ground to pick up vibrations, transmitting them through its skull bones to the ear ossicle. An erect dinosaur, towering ten or twenty feet above the ground, had totally lost this ability and relied solely on sound waves travelling through the air (a far less efficient means of hearing – especially when only one bony ear ossicle is present, as is the case with reptiles : in mammals there are three to amplify the vibrations). With the ever-present danger of giant tyrannosaurs, the duck-bill had to equip itself with some equally efficient alternative sense if it were to remain as vigilant. Not only was its sense of smell acute but its eyesight was also excellent and standing upright on its hind limbs served both to enable the animal to sniff the wind and scout the horizon for danger. These peaceful yet vulnerable dinosaurs were constantly wary to the presence of danger and had augmented the biped's

16. The Canadian duck-billed hadrosaur *Hypacrosaurus* was 25 feet long and walked, like most bipedal dinosaurs, with tail outstretched.

good eyesight with an equally good sense of smell to keep one jump ahead of the really giant carnivores of the day. This new line of reasoning raises an interesting point: what produced the characteristic smells of a dinosaur, if, indeed, they produced any at all of significance? The scent of sweat and body oils in a mammal carries far and can quickly be detected with a keen sense of smell. What of dinosaurs? Did *Tyrannosaurus* produce oils to lubricate its leathery skin that would unintentionally serve to alert plant-eating dinosaurs? If it did not we can only presume that the hadrosaurs' acute sense of smell functioned to detect the more aromatic tropical fruits, or, like an elephant's keen sense of smell, was used to detect the presence of water.

Ostrom went on to strengthen the theory by totally removing these duck-bill dinosaurs from their supposed swampy homes and placing them out on dry land where, of course, the tyrannosaurs reigned supreme. In this environment a good sense of smell would have been a definite asset. Ostrom produced many arguments in support of his thesis. The localities where hadrosaur skeletons are frequently unearthed sport abundant terrestrial plant remains but very few aquatic plants. Hadrosaurs lived on low-lying coastal plains, permeated by broad meandering rivers and covered by extensive coniferous forests. The hadrosaur dental structure was perhaps the most specialised of all dinosaurs' and was completely suited to abrasive woody plant matter. In the beaked hadrosaur jaw there was a complex battery of interlocking grinders: five hundred teeth in each jaw composed the grinding plates, so there were as many as *two thousand* teeth in the mouth altogether. The prism-shaped teeth were compressed together into a grinding mill with an uneven crushing surface. These teeth always show a high degree of wear and the fact that they were continually replaced from below when worn out also suggests that the food was highly abrasive. It was certainly not the succulent and soft aquatic weeds that had first been suggested by the animal's duck-like bill. An elaborate masticatory action of the jaws was perfected to grind plant material between the dental batteries. Unlike the cow's side-to-side cutting motion, the jaws moved fore and aft upon one another, rather as they do among the present-day rodents. Ostrom's suspicion that hadrosaurs were land-dwellers and browsed off conifer trees was justified when he rediscovered a paper written in 1922 on the stomach contents of a mummified *Anatosaurus* (or 'duck reptile', so called on account of its hadrosaurian beak). Besides an abundance of conifer needles, the animal had consumed twigs, fruit and seeds. Since these facts came to light during the heyday of the aquatic hadrosaur, the results were lightly glossed over at the time. In 1964 Ostrom refocused attention on to these findings as evidence of the most convincing kind that hadrosaurs were terrestrial browsers. Multitudes of these beasts undoubtedly spent their days stretching up on their hind legs to crop the conifer and poplar trees in the extensive Cretaceous coastal forests.[33]

The explosion of tall upright hadrosaurs in the later Cretaceous called into existence a host of predators, all of which likewise walked on their hind legs. In the bipedal plant-eaters, though, the forelimbs were never reduced to the extent seen in carnivores. Like living ungulates (hoofed mammals), many dinosaurian herbivores bore hooves, which tends to suggest that they could easily resort to all fours,[34] rearing up on to their hind legs perhaps to sniff the wind or reach the tallest leaves.

Unlike a large hadrosaur, which 'only' weighed as much as a small elephant (about 3 tons), *Tyrannosaurus* may have exceeded 8 tons and was consequently constructed far more rigidly. The rear legs were stout. The pelvic girdle was broad, allowing extensive limb muscle attachment, and fused solidly to the backbone to take the strain when the animal stood. It was at this point that the body was pivoted and the entire 8 ton weight was directed solely through the rear legs. For many years the front limbs and hands eluded the American Museum teams searching in Montana. When traces of these were eventually retrieved, they proved to be so ridiculously stunted as to suggest that the front legs were totally useless. No one could conceive of a function for such vestiges.

The long powerfully constructed walking legs raised the beast's hip region ten feet off the ground; its knee would tower above a human head. By contrast, the feeble arms were barely thirty inches long. The front legs in the older Jurassic carnivore *Allosaurus* were appreciably longer and sturdier, although in this case too they were totally useless for walking and must have functioned solely in fighting or tearing open the prey. At least, large and very strong claws are suggestive of offence.

The early restorers of *Tyrannosaurus* portrayed it with forelimbs terminated by the customary three fingers. Some decades later, when the front limb eventually turned up, it was found rather surprisingly to possess only two fingers, both clawed. The limbs were indeed puny and so short that they could not even reach the beast's mouth. This great flesh-eater, the largest land-based carnivore in earth history, had apparently dispensed with its forelimbs, abandoning the function of killing its giant contemporaries to its massive jaws and rear talons.[35] Thus the last carnivore of its kind had advanced the trend already seen in its allosaur and megalosaur forerunners to its fullest expression. The tyrannosaur was totally incapable of a normal quadrupedal pose and was forced to balance on its enormous rear legs; rocking on the pelvic girdle, it employed its giant tail to offset the weight of the head and trunk as it walked. Presumably it would recline on occasions, resting on its haunches and employing its tail as a prop. The slightly older *Gorgosaurus* shows a close similarity to *Tyrannosaurus*, especially in its possession of only two fingers in the hand. Lambe depicted this beast walking with a definite stoop rather than adopting the imposing upright stance always accorded to *Tyrannosaurus*. He also found in the leg joints signs that it had been bow-legged in life: a stance hardly becoming its majestic station.[36] Although most economical under the conditions of tremendous size and weight, *Gorgosaurus'* stooping walk with tail raised off the ground as a counterbalance must have been exhausting. The near loss of the front legs in *Gorgosaurus* and *Tyrannosaurus* demonstrates their irrevocable commitment to a two-footed pose. But were the vestigial front legs as useless as is generally assumed? Strangely, the diminutive limbs are adorned with relatively large claws, which is inexplicable if they were indeed functionless. Even more surprisingly, the bracing pectoral girdle was large, indicating an arm well

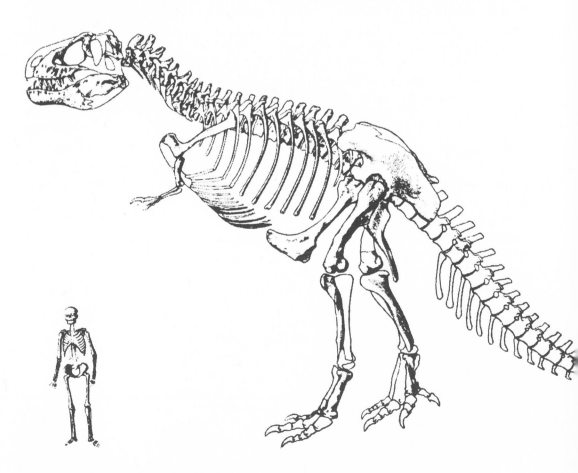

endowed with muscles. This is highly suggestive: the limb could not be as completely useless as was previously thought.

In a radical reinterpretation of *Tyrannosaurus*, Barney Newman at the British Museum recently suggested that the forelimbs did indeed have a role to play in the beast's life. Although the forelimb refurbished with flesh was only as thick as a man's thigh, it was strong enough to help raise the tyrannosaur off the ground after resting. A squatting eight-ton tyrannosaur tucked its hindlimbs under its body chicken-fashion. With tiny front legs stretched forward, it laid its head down so the jaws rested on the ground, rather as a cat does occasionally. When the beast aroused, any attempt to raise the hindquarters by extending the rear legs would merely have caused the head to slide along the ground. This is where the front legs helped. The flexed claws dug in to redirect the efforts of the hind legs into raising the hindquarters rather than pushing the animal forward. Like a cow, the tyrannosaur then struggled to push its still prostrate forequarters off the ground. Small forelimb size meant that the tyrannosaur was forced to throw its head back while straightening its legs. With the hind limbs fully extended, the tyrannosaur could wander off in search of prey. Like *Gorgosaurus*, *Tyrannosaurus* walked with a stoop in Newman's new picture. He suggests that the body was held horizontally rather than inclined at 45° as in conventional restorations. The flexible vertebral articulations allowed a 'swan-neck' curve to the neck and permitted the manoeuvrability so essential to such a killing machine. Newman even lopped twelve feet off the tyrannosaur's tail to produce a more symmetrical creature. Actually, this is not as arbitrary as it sounds, since the existence of this last twelve feet was only conjectured, there are no specimens in existence in which it can actually be seen. *Tyrannosaurus* now has a more balanced weight distribution fore and aft. Rather than dragging its tail along the ground after it, where it would have failed in its function as a counterpoise, *Tyrannosaurus* raised it off the ground to lessen the effort of bearing the trunk. The graceful motion that was once accorded to the king of killers was dealt a severe blow by Newman's study of its limb joints. When the leg was swung forward tendons stretching to the toes automatically tightened and the toes clenched like those of a bird on a perch. This allowed the foot to clear the ground. As the free leg advanced it swung the enormous tail to one side, producing an ungainly duck-like waddling in the monster. One megalosaurian trackway from Swanage now in the British Museum shows a sinuous trail with toes pointing inwards. Like the megalosaur, *Tyrannosaurus* was probably pigeon-toed.[37]

17. The first sketch of *Tyrannosaurus rex*, the largest terrestrial carnivore in earth history, made in 1906.

18. According to Newman, *Tyrannosaurus* used its apparently vestigial forelimbs as a brace to stop the head sliding along the ground as it rose.

It is immediately appreciable from one glance at the waddling *Tyrannosaurus* that the effort in just staying on its feet must have been enormous. The mere act of balancing an 8-ton frame across the hind legs must have utilised a great deal of energy, energy spent by the muscles of its legs, tail and vertebrae in countering the great gravitational force. Unlike the lizard, which runs for short spurts before flopping on to its belly to recuperate, *Tyrannosaurus* did not sprawl with its belly just off the ground but walked with legs tucked under the body and belly raised high into the air. It was not built for flopping on to the ground. Usually, it was forced to remain balanced on its feet. This is just the situation we encounter in a mammal and bird. Anything from half to three-quarters of a mammal's or bird's active life is spent standing upright. The same was probably true of *Tyran-nosaurus* (and all dinosaurs). Some degree of energy-saving could undoubtedly be achieved by the tyrannosaur reclining on its haunches. Even this leaves it in nothing like the total rest position of a prostrate lizard. The problem, of course, is greatly magnified when we consider a titanic brontosaur: forty or fifty tons of flesh destined to remain on its feet for much of its waking life in order to continue searching for food. The energy needed must have been enormous. The energy spent by the muscles to hold fast the limb joints and backbone and thus keep the animal on its feet must not be under-estimated. Muscles must constantly be in tone just to stop the skeleton from collapsing; the greater the weight of the animal the greater the exertion. A lizard has relatively little to worry about since it spends very little of its life standing, rather it saves energy by collapsing on the ground. It has to, *it does not have the energy to remain standing*. It is forced to spend anything up to 90% of its 'active' life lying motionless!

John Ostrom was the first to fully grasp the implications of this, implications which demanded a complete overthrow of conventional ideas. In 1969 he spoke out against the accepted view of the dinosaur as a cold-blooded and sluggish reptile. The occasion was, appropriately, the first North American Paleontological Convention, held in Chicago's Field Museum of Natural History in 1969. Ostrom's chosen topic was palaeoclimatic inference from contemporary fossils. Previously, it had been assumed almost uncritically that the presence of the giant cold-blooded dinosaurs – dependent as they were on the sun – in the high latitudes implied a uniformly tropical climate over a large section of the earth in Mesozoic times. Reptiles and amphibians, it is true, are good thermal indicators, but Ostrom refused to employ dinosaurs to this end. Mammals and birds, pointed out Ostrom, had high energy expenditures. A rat, for example, burns ten times as much fuel per unit volume as an alligator; hummingbird muscle uses over five hundred times as much oxygen as frog muscle of the same size, even though both are in the resting condition. These figures graphically demonstrate the amount of oxygen that is consumed to produce the vast quantity of energy needed by mammals and birds, energy that in part goes to maintain a consistently high body temperature internally. With high metabolic rate and fast energy production mammals and birds are able to stand and walk upright for almost all of their waking life. To assist them, the limbs have been moved under the body so they can stand erect *over* their legs, rather than slung between them as in a lizard.

So a lizard spends most of its waking life passively, whereas mammals and birds really do spend it actively, standing, walking, cleaning, prey catching, ritualising, and so on. 'The correlation of high body temperature, . . . high metabolism, and erect posture and locomotion is not accidental,' said Ostrom. 'The evidence indicates that erect posture and locomotion probably are not possible without high metabolism and high uniform temperature.'[38] And the dinosaurs *were* erect: herds of rhinoceros-postured *Triceratops* were stalked on Mesozoic plains by equally giant two-footed meat-eaters. These dominant Mesozoic reptiles had shared in the mammals' secret of success. They too were warm-blooded. As a corollary, the wandering of the dinosaurs into the cooler northerly latitudes, which they could combat with their own heat supply, makes them practically useless as thermal indicators. The wisdom in Ostrom's refusal to employ dinosaurs as indicators of palaeoclimate became apparent in 1973, when Dale Russell of the Canadian National Museums announced the discovery of duck-billed hadrosaurs *inside the Canadian Cretaceous Arctic Circle*.[39] Clearly, dinosaurs were not restricted to the equable latitudes frequented by sun-baskers; and Russell's discovery is itself strong supporting evidence that dinosaurs were far from traditional ectotherms. But even as endotherms, it seems improbable that hadrosaurs could have survived the dark, cold winter months at such latitudes, and Russell suggests that they migrated south to more hospitable regions at this time.

Nobody before had demonstrated the inextricable relationship between high metabolism, stable temperature and erect posture, yet once explicitly stated this linking seemed obvious and natural. It resolved the long-standing contradictions inherent in the ludicrous sun-basking brontosaur model by scrapping the model altogether and substituting an endothermic dinosaur. Of course, this requires a radical reappraisal of dinosaurian physiology; and we are compelled to look to the mammal and bird for our new model. The lizard's lungs are capable of extracting far less oxygen from the air than those of a mammal, where the lung tissues are made spongy by myriads of minute ultra-thin walled vesicles designed to expose a large surface area to the air. In a lizard these air pockets are much larger and the surface area is correspondingly diminished. Hence the blood receives less oxygen, a deficiency compounded still further, as we have seen, by a less efficient heart. The lizard heart has only a single ventricle, which serves to pump both the blood returning from the body to the lungs and that returning from the lungs back to the body. Since the blood from these two systems has to pass through a single chamber it gets mixed: so the oxygen-rich blood arriving from the lungs becomes adulterated with the spent venous blood before being pumped to the muscles where the oxygen is needed. The mammal and bird have split the ventricle into two chambers, one to cope specifically with the spent blood received from the tissues, which it pumps to the lungs; the other concentrates on sending only unadulterated oxygenated blood received from the lungs back to the body. This provides the large quantity of essential oxygen needed by a fast-metabolising, warm-blooded creature. It must also have been the system employed by the dinosaur. Indeed, the crocodile, which may be considered a frozen relic of the dinosaur's ancestor, has a functionally four-chambered heart, with a ventricular septum dividing the arterial and venous systems.

The dinosaur's high metabolism and fast energy production places it not with the cold-blooded lizards but the warm-blooded mammals and birds. Richard Owen would have been both surprised and delighted. Surprised that a reptile not only had some mammalian attributes, but was even warm-blooded; and delighted by the spectacular justification of his belief that the dinosaur really was the crown of reptilian creation.

3. The race is to the swift, the battle to the strong

This still highly controversial idea began at once to claim supporters, who saw in it one of the most fruitful and productive of concepts. It suggested new questions to ask and opened up new lines of approach which unexpectedly revealed the dinosaur in a wholly new light.

By a remarkable coincidence, the new light had simultaneously flickered on the other side of the Atlantic. Armand de Ricqlès at Paris University had arrived *independently* at the same conclusion within months of Ostrom, although his avenue of approach had been quite different. (This sort of happy coincidence is a recurrent feature of scientific discovery. At times, novel ideas seem to hang in the air, awaiting those with the courage to pluck them; recall the battle between Newton and Leibniz over who invented the calculus, or Darwin's shock upon receiving Wallace's abstract of his own theory of evolution by natural selection.) So in 1969, after his study of the internal structure of many types of fossil and recent bone, Ricqlès suggested that dinosaurs were more akin to mammals than lizards in their physiology. Bone, besides its obvious role as internal scaffolding, is an active, growing and physiologically important structure: blood cells are manufactured in the marrow, and bone tissue acts as a major calcium store (hormones trigger the release of the bone's calcium and dispatch it to the sites of muscular contraction, where it plays an important role). An energetic animal, a fast-metabolising beast with its physiological systems speeded up, will make a greater demand on its bone, and this will show in its fine structure. Ricqlès realised that bone structure was a sensitive indicator of an animal's level of activity, and that in fossil bone we had all the ingredients to conjure up the physiology of long dead animals. In mammals and birds the blood vessels permeating the long bones are numerous and densely packed (allowing a greater transport of materials to and from active regions); in reptiles blood vessels are sparse. The Haversian canals, the sites controlling the exchange of calcium between skeleton and blood, are likewise less numerous in reptile bone than mammal bone, and their lack restricts the speed of exchange. Ricqlès emphasised that, *on both counts, dinosaur bone resembles mammal bone*, but was sharply differentiated from the bone of cold-blooded reptiles and amphibians. He interpreted this striking 'convergence' as evidence of an active metabolism, and perhaps warm-bloodedness, in dinosaurs.[1]

These independent studies by Ostrom and Ricqlès furnished the mammal as the new paradigm for the dinosaur and thus removed the strictures associated with the obsolete lizard model. Switching models focused attention on to

19. Lawrence Lambe pictured *Gorgosaurus* as a slothful beast, spurred into motion only by the pangs of hunger.

previously unexplored aspects of the dinosaur, aspects that would never have suggested themselves in the old order. It was in the hands of Ostrom's student at Yale, Robert T. Bakker, that these novel ideas were most fully explored. Bakker took these radical views with him to the Museum of Comparative Zoology at Harvard University and it was here, in Louis Agassiz' museum, that he sought out the implications of endothermy for dinosaurs in a series of papers in the early 1970s. These new lines of approach also recast existing problems in a new perspective. Often such problems had passed unrecognised and paradoxes were happily accepted without embarrassment. The lizard's physiological handicaps could not be ignored, and though it was conceded that living reptiles are hampered by their limited ability to run or even stand upright, dinosaurs were assumed to have performed prodigious feats of strength and endurance. Even though the limitations would have increased out of all proportion if the lizard were elephant-sized, still the formidable meat-eaters were supposed to have spent their lives in active pursuit of prey, whilst the plant-eaters in their turn were believed capable of fleeing from their clutches. If a small lizard weighing a few ounces can only sustain rapid bursts of activity, how could an eight-ton flesh-eater like *Tyrannosaurus* ever have endured a protracted gladiatorial struggle with its prey? Death for *Triceratops* in the jaws of the tyrannosaur could not have been as instantaneous as it is for a fly snapped up by a lizard. A large dinosaur had about a million times the bulk of a small lizard, yet it was presumed to have chased its prey and engaged in mortal combat. Where did it find the energy?

One way out of of the paradox was to assume, as some did of the huge flesh-eaters, that they were too large and ponderous to have actively stalked prey and then have done battle. Lawrence Lambe, the original describer of the Canadian

63

monster *Gorgosaurus*, was the most ardent supporter of this view. Locating a new dinosaurian carnivore is always something of an event. Flesh-eaters were not nearly so abundant as their inoffensive herbivorous contemporaries in Mesozoic times, but being such imposing beasts the deficiency in numbers is quickly made up for in the discoverer's eyes. Often a community in one locality will support a variety of herbivores but only a single giant carnivore to keep them in check. Opening up the Canadian Red Deer River region of Alberta early in the century thus brought to light many new plant-eaters, and in 1913 placed into Lambe's possession an entirely new carnivore. *Gorgosaurus* was a late Cretaceous giant: full length it attained 29 feet and stood 11 feet tall. Armed with the obligatory stout teeth and claws, the gorgosaur was nevertheless of a lighter build than *Tyrannosaurus*, with slimmer and more graceful hind limbs. Yet it obviously weighed some tons and Lambe thought that the creature must have spent most of its waking life prostrate on the ground. It could walk erect if the situation demanded, when the tail acted as a counterpoise to the trunk and the full weight was borne by the hind legs, but this was a rare event called forth only by the strongest incentive. In a prone position, sitting or squatting, the weight of the gorgosaur was passed directly through the pelvic girdle to the ground. 'This position of rest,' hazarded Lambe, 'and particularly the recumbent one of repose at full length [when the creature lay flat-out with its head on the ground] were probably those most frequently assumed by a reptile having the form, and the supposed sluggish disposition of Gorgosaurus.'[2] Strangely, the long serrated teeth in the gorgosaur's jaws were completely unworn, yet the individual was not a juvenile. The tips of the fangs were almost perfect, even retaining delicate serrations on the sharp cutting edges. Lambe inferred that the food taken by this meat-eater was soft and unabrasive, and that no struggle could have been put up by the prey.

> It is believed, therefore, that Gorgosaurus confined itself to feeding upon carcases of animals that had not been freshly killed, that it was not as an intrepid hunter but as a scavenger that it played its useful part in nature, and no doubt its services were fully required when we consider the immense numbers of trachodonts, ceratopsians, stegosaurs, and other dinosaurs and reptiles that lived and died in this particular time of the Cretaceous period.

Engulfed as Lambe believed it was by slowly moving mounds of potential food, *Gorgosaurus* had no need of active predation. All manner of plant-eaters must have died naturally in its vicinity. The beast gorged 'on carcasses found or stumbled across during its hunger impelled wanderings'. The monstrous gorgosaur was a near-harmless carrion-eater. It would tear the soft tissues of the corpse with its claws before squatting over the rotting flesh and consuming the palatable tissues. Or so Lambe thought.

Unfortunately for Lambe's theory, teeth are lost or worn, not during the actual kill, but in tearing meat from the bones. The American Museum has a brontosaur skeleton complete with tail bones that have been gnawed by an allosaur. When an allosaur jaw was compared with the score marks there was an exact fit; an

20. *Allosaurus*, in the American Museum display, hovers over a brontosaur backbone. The brontosaur vertebrae were unearthed bearing allosaur tooth marks.

Allosaurus had obviously feasted on the brontosaur carcass. Moreover, in the process it had lost several teeth, and these were found lying beside the brontosaur tail when it was extracted from the rocks.[3] Why the Canadian gorgosaur's teeth are unworn is puzzling since merely feasting on a carcass would have worn them down. Certainly, this cannot be used as evidence to prove the gorgosaur's cowardly behaviour. The battery of armaments it sported were just as fearsome as those of the allosaur or tyrannosaur. Scavenging seems an unlikely pre-occupation for this imposing beast; it *looks* as though it were built for active aggression. Moreover, most investigators concerned with large carnivores like *Allosaurus* and *Tyrannosaurus* from North America and *Tarbosaurus*, the tyrannosaur's Mongolian counterpart, failed to share Lambe's scepticism, preferring to believe instead the evidence of the jaws and talons: these signified formidable killers. In life (and death), of course, things are never quite this clear cut – a lion will scavenge and a hyena kill on occasions. It would be unrealistic to imagine a gorgosaur or tyrannosaur passing up a free meal if they stumbled upon a corpse. Nevertheless, it seems more probable that like lions they were primarily active killers.

Tyrannosaurus rex, the 'king of the tyrant lizards', was the very last expression of its kind. Its impressive size, just under forty feet if we accept Newman's revised estimate, and standing nearly twenty feet tall on its rear legs, leaves no doubt that it was the Mesozoic king of beasts. It was first brought to light as a result of an intense period of activity at the American Museum at about the turn of the century. Under the Presidency of Henry Fairfield Osborn, a great many well-equipped expeditions were dispatched to the Rocky Mountains, which resulted in such prolific discoveries that we are still reaping the benefit. *Tyrannosaurus* emerged from its Hell Creek grave in Montana in 1902, when Barnum Brown's dinosaur-hunting team scoured the region. Brown, accompanied by Lull, located the beast's skull, jaws, backbone and hind limbs, but the

21. The Canadian *Gorgosaurus* eyeing a small armoured ankylosaur.

hard entombing Upper Cretaceous sandstones forced the museum crews to return two or three years in succession to completely disinter it, while back at the museum Lull prepared the skeleton for mounting. *Tyrannosaurus* took everybody by surprise. *Allosaurus*, large as it was, was dwarfed by this greatest tyrant of all. Osborn's colleague at the museum, William Diller Matthew, attempted a preliminary sketch of the beast and to emphasise its dimensions in the most dramatic fashion stood it next to the diminutive human skeleton. This specimen and much of two additional skeletons found on the successive forays to Hell Creek were studied by Osborn, and he christened the creature *Tyrannosaurus rex* in recognition of it having been the largest land-living carnivore ever to have stalked the earth.[4] A creature, furthermore, that was constructed solidly and undeniably for battle, with lethal teeth and a frightening gape, and eight-inch talons arming its toes. Just as Cope, upon finding the carnivorous *Laelaps* in the same area as the hadrosaur, had felt compelled to set them upon one another, so Osborn suggested that the tyrannosaur was the relentless enemy of the horned *Triceratops*. The armour plating of the three-horned herbivore had reached such thickness that *Tyrannosaurus* had to evolve six-inch teeth to penetrate it, and a jaw four feet long to manage the chunks of flesh.

One glance at the skull of *Tyrannosaurus* reveals its predatory habits. Since the forelimbs were useless in rending the flesh of the struggling prey, the skull had

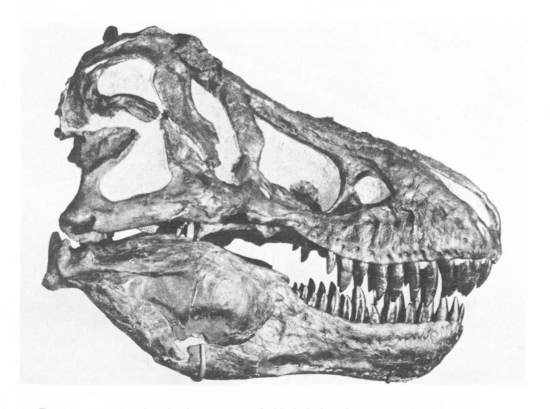

22. *Tyrannosaurus*, whose four-foot jaws were armed with six-inch teeth, was unquestionably a ruthless culling agent which actively killed its prey.

assumed the main function of offence. Consequently it had grown disproportionately large for the animal's size and was armed with a battery of curved teeth serrated like palaeolithic spear points. The length of the jaws allowed a really effective bite and *Tyrannosaurus* was able to swallow sizeable masses of dinosaur meat. The smaller *Allosaurus* surpassed even the tyrannosaur in its ability to gulp down unwieldy chunks of prey whole. In the temple region of the skull the quadrate bone, which articulated with the lower jaw, was movable to a small degree, whilst both the jaw itself and the skull roof could flex at their mid points, a series of remarkable adaptations that enabled the gape to stretch still further and the *Allosaurus* to greedily bolt enormous masses of flesh.

Two of the tyrannosaurs secured by Barnum Brown in Montana were restored in the Cretaceous Hall of the American Museum in New York. The poses that were adopted by Osborn for the reconstructions were those drawn up by the Curator of Reptiles at the New York Zoological Park after an intensive study of lizard movements. These restorations, said Osborn, depicted 'animals prior to the convulsive spring and tooth grip which distinguishes the combat of reptile[s].'[5]

Barnum Brown described the scene a little more vividly.

> It is early morning along the shore of a Cretaceous lake three millions of years ago. [In Barnum Brown's day – this was written in 1915 – far less time was allowed for the duration of earth history: it is now believed that 70 million years has elapsed since the close of the Cretaceous.] A herbivorous dinosaur *Trachodon* venturing from the water for a breakfast of succulent vegetation has been caught and partly devoured by a giant flesh-eating *Tyrannosaurus*. As this monster crouches over the carcass, busily dismembering it, another *Tyrannosaurus* is attracted to the scene. Approaching, it rises nearly to its full height to grapple the more fortunate hunter and dispute the prey. The crouching figure reluctantly stops eating and accepts the challenge, partly rising to spring at its adversary.
>
> The psychological moment of tense inertia before the combat was chosen to best show positions of the limbs and bodies, as well as to picture an incident in the life history of these giant reptiles.[6]

Cope had established the precedent by setting his *Laelaps* at each other's throats, so Brown follows suit and pits one tyrannosaur against another. But such intraspecific aggression seems unlikely. The fact that many authors have assumed that the limited mental powers of the dinosaurian predator made anything that moved fair prey is all the more reason to believe that some sort of (perhaps incipient) territorial behaviour was the rule amongst the larger flesh-eaters, as it is today amongst lions. The smaller predaceous dinosaurs may well have hunted in packs like wolves,[7] in which case there is very likely some sort of hierarchy – a dinosaurian peck order – to stop such mutually destructive brawls as Cope and Brown envisaged.

Brown thought *Tyrannosaurus* 'a powerful creature, active and swift of movement when occasion arose'. Indeed, after Brown himself had unearthed a skeleton of *Gorgosaurus* from the Red Deer River area of Alberta a decade later, he sent it back to the museum to be restored in a running pose, as if chasing a herd of duckbill dinosaurs.[8]

Osborn and Matthew, aware of the limited ability of the lizard to sustain active aggression, were more guarded in their statements concerning the aggressive struggles and the fleetness of foot of these multi-ton carnivores. The lizard makes a swift kill, one lunge and the insect's fate is sealed, and Matthew envisaged the same in the ancient world. 'As for its probable habits,' he said of *Allosaurus* (whose name means 'leaping reptile' – a holdover from Cope's time), 'it is safe to infer that it was predaceous, active and powerful, and adapted to terrestrial life. Its methods of attack and combat must have been more like those of modern reptiles than the more intelligent methods of the mammalian carnivores.' He imagined *Allosaurus* lying in wait (surely one of the intelligent methods of today's sophisticated predators?), 'then a sudden swift rush, a fierce snap of the huge jaws and a savage attack with teeth and claws until the victim is torn to pieces or swallowed whole.'[9] It was as if, in his eyes, an eight-ton carnivore could mimic a lizard snapping at its prey and hastily retiring to regain its strength. This, of course, is highly improbable. The massive bulk of *Allosaurus* and its foe would have militated against any such snap victory. The prey was not a minute inoffensive creature, as is a lizard's prey, but for *Tyrannosaurus* it was supposedly an armoured three-horned ceratopsian or a more agile duckbilled hadrosaur.

Surprisingly, Matthew contradicted himself when he came to review the probable habits of the tyrannosaur. 'Its bulk precluded quickness and agility,' he declared, oblivious that that statement must also have been applicable to the *Allosaurus*, which itself attained thirty-five feet in length. 'It must have been designed to attack and prey upon the ponderous and slow-moving Horned and Armoured Dinosaurs with which its remains are found, and whose massive cuirass and weapons of defence are well matched with its teeth and claws. The momentum of its huge body involved a seemingly slow and lumbering action, an inertia of its movements, difficult to start and difficult to shift or to stop. Such movements are widely different from the agile swiftness which we naturally associate with a beast of prey.' No animal of tyrannosaur dimensions could leap or spring upon its prey, he concluded, and a swift advance into battle may well have ended in unavoidable impalement on the sharp horns of *Triceratops*. How, then, did Matthew picture the 'combats of titans of the ancient world'? Presumably they were protracted struggles against well-armoured adversaries, struggles leaving little respite for the protagonists to regain their strength. Could a cold-blooded, physiologically inefficient, eight-ton reptile have endured this with its low energy producing metabolism? Would not an ectothermic tyrannosaur, in point of fact, have been forced to retire *more* promptly than a lizard, whose diminutive weight, after all, subjects the muscles to far less strain?

W. E. Swinton, discussing the habits of the tyrannosaurs in 1934, clearly envisaged a lengthy gladiatorial clash during the primeval reptilian kill. '*Tyrannosaurus*,' he said of the Cretaceous super-predator, 'could not have been other than a clumsy and awkward giant battling against other equally cumbrous forms.'

Not a cunning or highly-brained creature, it would be guided largely by instinct and automatic reactions to the stimuli of sight and smell, even though long practice through countless carnivorous ancestors had brought this co-ordination to a high pitch of efficiency. Still, the mechanical limits of the body were all against sudden leaping movements, swift pursuit, or the battle of wits that characterizes mammalian contests; and, no doubt, the contests of that Cretaceous world, could we see them now, would seem to have the stiffness of amateur activity rather than the smoothness of the professional even though the feud was real and terrible.[10]

Inferring the function of structures (such as teeth and claws) from their shape is only one of the means, albeit the main one, of demonstrating that the theropod (flesh-eating) dinosaurs were active aggressors. During the late 1930s, Roland T. Bird, a dinosaur scout for the American Museum and fellow collector with Barnum Brown, caused quite a stir with *his* dinosaur discoveries. Well, not exactly dinosaurs themselves so much as their tracks, left for posterity when the ancient leviathans moved across mud flats. On those occasions when the mud was sufficiently stiff, well-defined tracks were left which filled up with sediments washed across the flats. When the land was elevated and the overlying rocks eroded millions of years later this shale inside the tracks was weathered out more easily to leave near perfect impressions of the ancient footprints. The science dealing with fossil footprints, or ichnology, was given an immeasurable boost by Bird's numerous impressive finds in the Rocky Mountain states.

Towards the end of November 1938 Bird was led to the scene of some outsize three-toed carnivore's tracks – perfect right down to the friction pads – after seeing some track-bearing slabs of rock for sale in an Indian trader's store. Three-toed dinosaur tracks were not exactly rare, many had come to light in the century since Edward Hitchock collected his 'bird' tracks from the Connecticut valley for his museum at Amherst College. Hitchcock had uncovered the tracks of primitive dinosaurs that inhabited New England in Triassic times, whereas Bird was dealing with much younger rocks, about 120 million years old, at Glen Rose, 80 miles south-west of Fort Worth in Texas. Glen Rose was virgin territory for the American Museum expeditions. That the region was going to be profitable was obvious to Bird from the block of masonry impressed with a 20 inch print of a carnivore that had actually been built into the local country courthouse. The town, it seems, had long taken such 'man tracks' for granted. They occurred in large numbers on the bed of the local Palaxy River some miles upstream where the river was cutting through Lower Cretaceous mudflats and exposing the footprints. Bird spent the last days of that season excavating these eagle-clawed carnivore tracks, but he also uncovered, almost by accident, the first known four-toed tracks of the lumbering brontosaur giants. These 38 inch sauropod prints sent him scurrying back to Glen Rose the following season. Work teams from the American Museum and the University of Texas quarried some 40 tons of track-bearing slabs from the site, in the process of which they stumbled upon one particularly spectacular trail. A stately sauropod had crossed the mud flats unaware that a flesh-eater was following close behind. When the

23. Shortly after locating the first sauropod tracks in Glen Rose in 1938, Roland T. Bird uncovered this spectacular trail, which he interpreted as a brontosaur being pursued by a carnivore. The 3 foot, 18 gallon, brontosaur prints are flanked to the left by the three-toed prints of an *Allosaurus*-like flesh-eater.

brontosaur swung to the left, the tall carnivore followed. Tension mounted in even the most stolid workmen as the drama unfolded and more of the tracks were brought to light. There was even a certain amount of friendly rivalry in unearthing the footprints, and speculations on the outcome of the prehistoric chase. Would the brontosaur flee to the safety of the deep, or would his half eaten skeleton be yet found somewhere ahead? The anticlimatic end of the pursuit came as the ledge ran under a limestone rock face. The outcome remains hidden.[11] Was there a chase in the first place? Most authorities seem to have accepted this as evidence of the most dramatic form that the brontosaurs were hounded and ultimately fell prey to the rapacious carnivores trailing them. Since, as Richard Swann Lull once said, footprints 'are fossils of *living beings*, while all of the other relics are those of the dead', Bird's evidence counted for a great deal.[12] He presented to the scientific world a brief episode in the *life* of one particular brontosaur; he focused on one day, a day that may have witnessed the ending of its life.

There was plainly a certain amount of double-think going on. On the one hand today's reptiles were recognised as possessing severe handicaps when it came to sustaining any sort of activity, yet yesterday's glorified reptiles were capable of the most extraordinary feats. The paradox was only recognised as such by a few and never explicitly stated, nevertheless it was implicitly accepted by all. The outcome of the struggle may have remained hidden, but there was very little doubt that there *had* been a struggle. The meat-eaters had not been the cowardly scavengers that Lambe supposed so much as cold-blooded murderers who terrified their victims.

The outcome of one battle is known. In August 1971 the joint Polish-Mongolian palaeontological expedition to the Gobi Desert came upon the fossilised skeletons of two dinosaurs which had seemingly perished whilst actually locked in mortal combat. The predator, a small agile *Velociraptor* with slender legs and long clenching hands (hence its name, which means 'swift robber'), was found grasping the skull of a small armoured *Protoceratops*, inextricably locked around the herbivore in a pose suggesting that both had died in the struggle.[13] (How two combatants could have died simultaneously and then have been preserved as if frozen in battle is a little hard to imagine.)

The paradox of the dinosaur as an active aggressor was neatly resolved by Bakker in 1972, when he examined the ecology of dinosaurian communities. The ratio of predators to prey in reptilian communities differs significantly from that in mammalian ones, a situation growing out of the differing energy requirements of the two types of creatures. The largest living lizard, the Komodo dragon (in fact a monitor closely related to the marine mosasaur of the Cretaceous), surviving on the remote island of Komodo east of Bali reaches twelve feet long and over 100 lbs in weight. This lizard will scavenge the equivalent of a pig's carcass about once a month. Since this is approximately half its own weight, it consumes its own weight in prey every sixty days. A cheetah, on the other hand, is required to burn a larger quantity of food to meet its higher energy requirements. It will take its own weight in food in only ten days. Lions consume even more, their own weight in as little as eight days, whilst in wild dogs it may be less than a week. Warm-blooded creatures therefore require anything up to ten times the quantity of food taken by a lizard. How are we able to apply these findings to the dinosaur communities of 70 million years ago? A community of creatures will only support a certain number of predators if the population is to remain stable. Because warm-blooded flesh-eaters make frequent kills, the population can only support few carnivores in relation to the inoffensive plant-eaters, so as not to annihilate them. In reptilian communities, since the killers take a lesser toll, more can be supported without diminishing the population. Bakker showed that in the fossil reptile communities of the mid-Permian there were far more predators than in fossil mammal communities, such as those of the Oligocene and Pliocene. What of dinosaurs? They showed a pattern analogous to the *mammalian* predator-prey ratios. As on African plains today, where antelope overwhelmingly outnumber lions, so in the late Cretaceous duck-billed hadrosaurs and horned ceratopsians were far in excess of the contemporary tyrannosaurs. Bakker's calculations on energy flow in dinosaurian communities makes it pretty certain that the low carnivore to herbivore ratio was due to warm-bloodedness and not to some quirk of community structure or even selective preservation. 'Analysis of energy flow strongly indicates,' concluded Bakker, 'that dinosaur energy budgets were like those of large mammals, *not elephant-size lizards*.'[14]

The really big killers in late Mesozoic times, the tyrannosaurs and allosaurs, were obviously not as strongly in evidence (at least in terms of numbers) as one would have expected if they were monstrous lizards. Nonetheless, being highly active, they were the rapacious beasts that one has been taught to believe and were required to kill frequently. *Tyrannosaurus*, at least, with its gaping jaws and eagle-clawed feet, had the capacity to bring down a large duckbill or ceratopsian. And it seems highly probable that these were its prey. It is unlikely that it could have caught one of the smaller, more nimble dinosaurs. The tyrannosaur's battery of armaments were designed to deal effectively with the larger, if well defended, plant-eaters like *Triceratops*.

Not that *Tyrannosaurus* was unable to run; it was presumably forced to on occasions in order to catch the galloping *Triceratops*. The tyrannosaur may have waddled duck-fashion, but even ducks can put on a show of speed when necessary. As Barnum Brown suspected, both *Tyrannosaurus* and *Gorgosaurus* were undoubtedly fairly swift during the actual kill. The larger ceratopsians like the rhino-sized *Chasmosaurus* and larger ten-ton *Torosaurus* and *Triceratops* could certainly gallop at a respectable speed. Like rhinos, ceratopsians were armed with an array of horns on the skull (some of which may have approached four feet long in life: the bony core was covered with a pointed horny sheath that is lost in the fossil) and would probably only retreat if a preliminary charge failed to shake the aggressor. Bakker estimated from his study of dinosaur limb joints that the large horned dinosaurs could gallop at least as fast as a rhinoceros. 'Armed with long horns on a highly manoeuvrable head, strong beaks, and the ability to gallop at speeds probably up to 30 mph, these large ceratopsians must have been some of the most dangerous terrestrial herbivores ever to have

24. Bakker's researches on dinosaurian limb mechanics served to emphasise the similarity between dinosaurs and large mammals. The ceratopsian *Chasmosaurus* reached 17 feet in length and was remarkably rhinoceros-like, presumably sharing the charging disposition of the rhino.

evolved.'[15] Unlike the unmolested rhinos, ceratopsians had cause to be dangerous, for they shared their world with a terrible foe that could match them in strength.

Here we encounter a more serious example of the double-think bedevilling palaeontology since Victorian times. Leidy and Cope had no qualms about letting their 'kangaroo' dinosaurs bound after one another unchecked, and the notion of an active, on occasions fast-moving, dinosaur took root, and has been alluded to ever since. Indeed, the first genuine fossil dinosaur skeleton mounted in the United States was a hadrosaur in full flight. The problem had always been to attain a really life-like pose. 'In most kinds of construction,' said Charles Beecher of Yale University in 1902, perhaps referring to Hawkins' fated hadrosaurs, 'the concrete result is usually found to differ in many particulars from the ideal.' Hawkins' restorations had faded into historical curiosities, so

Beecher set about the rehabilitation of the dinosaur into our world anew. The snag was that with no living descendants there were no guidelines to the pose that should be given the dinosaur. What was needed, he urged, was a skeleton that had remained intact since the beast's death, locked in a position that it might have adopted in life. Luckily, Beecher's requirements were met by a superb specimen of the hadrosaur *Claosaurus* that had been found by Marsh's collectors in 1891. Its limbs were still articulated, and in particular the left femur had remained socketed to the pelvis. So the fact that the leg was directed forwards as if frozen midstrike settled the question of the pose to be given the mount. Beecher modelled the running hadrosaur after the posture of existing animals in motion, paying particular attention to photographs of the living *Chlamydosaurus* lizard, which will occasionally run for short spurts on its rear legs.

> An attempt has been made [said Beecher of the finished restoration of the running *Claosaurus*] to carry out this idea of rapid motion and to make all parts of the skeleton contribute to the completeness and realism of the general effect. In order to do this there must be the proper balance and the true swing of the living animal.
>
> It is intended, therefore, that this huge specimen, as now mounted, should convey to the observer the impression of the rapid rush of a Mesozoic brute. The head is thrown up and turned outward. The jaws are slightly separated. The fore arms are balancing the sway of the shoulders. The left hind leg is at the end of the forward stride and bears the entire weight of the animal. The right foot has completed a step and has just left the ground preparatory to the forward swing. The ponderous and powerful tail is lifted free and doubly curved so as to balance the weight and compensate for the swaying of the body and legs. The whole expression is one of action and the spectator with little effort may endow this creature with many of its living attributes.[16]

Likewise Barnum Brown two decades later restored *Gorgosaurus* at the American Museum in a running pose. This choice of gait is not surprising, even a cursory glance at dinosaurian limbs and joints should be evidence enough that many dinosaurs were built on the same lines as modern hoofed mammals. Ceratopsian limb joints, for example, were very like those of a rhino, as was the degree of forearm swing, leading inevitably to the conclusion that the two could have kept pace with one another.

The bipedal dinosaurs were often restored with legs in full stride. As it came to be realised that these 'upright' dinosaurs could travel very fast indeed, palaeontologists began to look at the mechanics of the limbs and the most efficient running pose. Recently, as we have seen, it has become commonplace to reconstruct these bipeds with a tendon stiffened outstretched tail acting as a counterpoise and a horizontal backbone, itself ensheathed in tendons in many dinosaurs to stop it sagging. Sheathing was of the utmost importance in reinforcing the backbone and tail; we are only just beginning to appreciate how essential it must have been. If dinosaurs walked with backs held horizontally, the vertebral column had to be strengthened to counteract the effects of gravity. In

Deinonychus, a carnivore little more than 8 feet long living in Montana in Early Cretaceous times – and without question the most extraordinary dinosaur to be found in recent years – this sheathing had been taken to an unprecedented extreme. The articulating facets of its tail vertebrae had grown into bizarre trailing rods reaching 18 inches in length, which encased the vertebrae to render the whole unit rigid as an aid to maintaining dynamic stability. Why the beast needed such an unusual stabiliser only became apparent when the foot was functionally analysed. John Ostrom, who discovered the dinosaur in 1964, had this bold claim to make when he placed it before the world in 1969:

> The foot of *Deinonychus* is perhaps the most revealing bit of anatomical evidence pertaining to dinosaurian habits and must have been anything but 'reptilian' in its behaviour, responses and way of life. It must have been a fleet-footed, highly predaceous, extremely agile and very active animal, sensitive to many stimuli and quick in its responses. These in turn indicate an unusual level of activity for a reptile and suggest an unusually high metabolic rate. The evidence for these lie chiefly, but not entirely, in the pes[foot].[17]

What was so staggering about the foot? *Deinonychus* was an obligatory biped, the shape of its arms and hands argues strongly against it ever having walked on all fours. The foot was distinctive in possessing only two toes that reached the ground. There was a shortened third toe, but this had become modified into an offensive structure bearing a lethal, 5 inch sickle-shaped claw. This toe had become specialised *exclusively* as a formidable weapon, whose function was unmistakably to gouge and tear. Yet to slash open its prey with this talon must have required an unparalleled sense of balance and equilibrium, *because the*

25. *Deinonychus* – perhaps the most extraordinary dinosaur to have been unearthed in recent years. Like other dromaeosaurids, it was surprisingly sophisticated. Reconstruction by Robert T. Bakker.

animal had to stand on one leg in the process, whilst simultaneously grappling with the struggling prey. Balancing precariously on one foot could probably only be accomplished so long as the beast had a rigid tail acting like an inbuilt tight-rope walker's balancing pole. The prey had to be held away from the body if it was to be killed by kicks from the hind foot. For this the arms were long and gangling and the hands better adapted for grasping and holding than in any other dinosaur. This enabled *Deinonychus* to grip the unfortunate victim and hold it at arm's length whilst the taloned foot disembowelled it. Since the scars on the vertebrae of this ferocious killer exactly matched those of the large flightless birds like emus and ostriches, Ostrom concluded that the backbone in *Deinonychus* must also have been held rigidly horizontal by ligaments. The long legs betray *Deinonychus* as a swift predator easily able to run its quarry to ground. This spectacular little creature, quite unlike anything found before (so different in fact that Ostrom is not sure where its affinities lie), emerged at an opportune moment. An earlier decade would have viewed it quite differently, totally missing the functional significance of its salient features. Coinciding in its re-appearance with the 'revolution' in our thinking, *Deinonychus* embodies all that is distinctly non-reptilian in dinosaurs. It gives us, so Ostrom claimed on introducing the beast, an entirely new insight into the surprisingly sophisticated capabilities of some of the rapacious flesh-eating dinosaurs.

In 1970 Galton similarly attacked the old-style restorations of bipedal herbivores, where the creature ran with its body inclined at an angle of 45°, with a limp tail trailing along the ground (where the drag would not only have slowed the beast considerably but would have caused unnecessary damage to the tail itself). Instead, he too tilted the biped forwards into a stoop; his new reconstruction of the flat-headed hadrosaur *Anatosaurus* 'in a hurry' bears little relationship to former attempts, but it certainly conveys the impression of speed, which was always lacking before.[18]

But until the late 1960s these creatures were considered basically reptilian; no matter how agile they looked, in life they were severely handicapped by lack of an adequate power supply. Appearances certainly deceived many palaeontologists, who found themselves having to ascribe a function to those long legs and settling naturally for a speedy escape when danger threatened. Yet even at optimum temperature (the time of greatest energy production) reptiles can only muster a tenth or twentieth of the energy available to mammals of similar weight. Thus a reptile weighing two pounds can *sustain* a speed of only 1 mph with the energy available to it, whereas a mammal can sustain 10 mph or more. The giant *Tyrannosaurus* could perhaps reach 3 or 4 mph if it were ectothermic and the *Anatosaurus*, even when hurrying, could do little better![19]

The larger and more sturdily built creatures do not demonstrate the absurdity of the situation so beautifully as do the long-limbed, long-necked, flightless-bird mimics among the dinosaurs – the coelurosaurs. It was universally accepted that coelurosaurs really were very agile creatures indeed: Colbert wrote of *Ornithomimus* in 1962 that 'the hind limbs are long and nicely adapted for fast running'.[20] *Ornithomimus*, the 'bird mimic', possessed many of the characters we

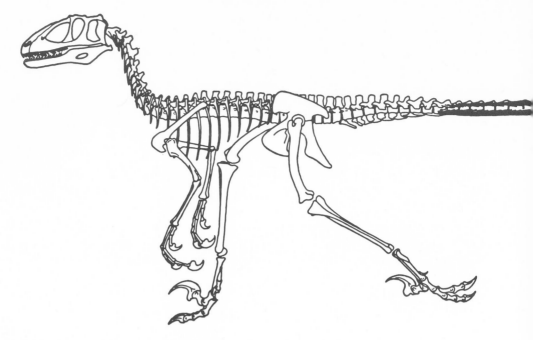

26. *Deinonychus*. Note the peculiar bony stiffening rods running the length of the tail.

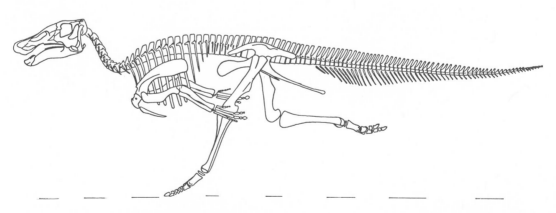

27. Peter Galton's duck-billed *Anatosaurus* 'in a hurry'.

now associate with the tall flightless running birds; it had delicate hollow bones, a long neck and small skull with a beak covered by a horny sheath in life, and very long legs. In the ostrich-mimicking *Struthiomimus* the last of the teeth had been lost which further enhanced the similarity to the ostrich. Like the ostriches *Struthiomimus* had an abbreviated back, brought about by shortening the distance between the shoulder and pelvic girdle, and the metatarsal bones of the

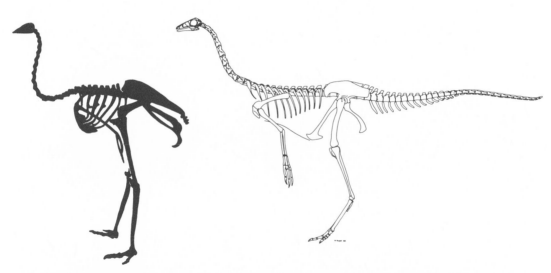

28. A modern skeletal reconstruction of the ostrich dinosaur *Struthiomimus* (right). Note the striking similarity to the ostrich (left).

foot had elongated to yield another functional segment to the leg, allowing it to run on its three toes (many fast creatures have done this to increase their stride). It was only the long tail, long arms and three-fingered hands that rendered the superficial similarity between *Struthiomimus* and the ostrich incomplete.

Struthiomimus was twelve feet long from beak to tip of tail and stood seven feet high, just a little taller than an ostrich. Since it possessed a toothless horny beak

29. Gerhard Heilmann's restoration of the ostrich dinosaur.

we have grounds for assuming that, although technically grouped with the theropodous flesh-eaters, the beast had switched to a herbivorous diet. The long delicate fingers were probably used to pull fruit from trees, or bend branches towards its horny beak.[21] When the struthiomimids were first recognised by Osborn in 1916, William K. Gregory of the American Museum suggested that the heavy bear-like claws terminating the fingers may have been used for tearing husks off fruit, or for plucking the fruits from palms, cycads and fig trees.[22] The weak toothless jaws could then cope with the soft fleshy fruit. Gerhard Heilmann was considerably impressed with Gregory's ideas and gave a remarkable restoration of a group of struthiomimids, suitably ostrich-like (too much so in many people's eyes[23]), feeding on such husked fruit.[24] *Struthiomimus* may not necessarily have been a strict vegetarian. The ostrich is omnivorous in the wild state and the struthiomimid may also have taken small lizards or mammals, catching them with its grasping hands (an advantage lost to the ostrich).

The powerful claws have been thought of as defensive weapons, but it seems more likely that the ostrich-mimic fled at the first sign of danger. Gregory had no doubt that *Struthiomimus* was a rapid runner when danger threatened. He even invested it with a flap of skin connecting the arms to the body that could act as an aerofoil to offset the weight of the doubled-over trunk as it ran. *Struthiomimus* really had no need of such aerodynamic assistance. The tail was a good three feet

80

longer than the trunk and was stiffened (as in other dinosaurs) in its posterior half to counteract the weight of the trunk. The marvellous convergence in shape between the ostrich and ostrich-dinosaur reveals much of the latter's habits to the functional anatomist. Convergence of appearance among animals is always accompanied by a close similarity of function. Creatures that look alike have similar habits, we have only to think of the marsupial sabre-tooth *Thylacosmilus* and the placental sabre-tooth 'tiger' *Smilodon*, ichthyosaurs and dolphins, and so on. Often related, sometimes starting from a totally different plan, these creatures have come to look like one another because they have come to *live* like one another. Thus if the ostrich has long leg bones to increase its stride and therefore speed, it is reasonable to assume that the same was true of *Struthiomimus*, that it too had lengthened its stride in order to increase its speed. Since ostriches can run at anything up to 50 mph, ostrich-dinosaurs could achieve the same speed: at least, they had the necessary mechanics for such a speed. But if they were only equipped with a reptilian powerhouse, if they possessed the physiology and energy output of a lizard, struthiomimids could have sustained a breath-taking speed approaching 2 mph![25] This paradoxical situation, where a dinosaur so obviously built for speed was restricted to a snail's pace, is tantamount to building the Concorde then equipping it with renovated Sopwith Camel engines!

Gregory was an expert on animal locomotion. Speed, as he showed, was related to the length of the stride and the rapidity of the step. To increase the length of the stride most fast mammals have increased the length of the bones in the distal part of the leg (the shin and foot region), and a comparison of the ratios of upper to lower leg bone lengths is a good indicator of the speed that an animal can attain. An elephant's femur (thigh) is much longer than its tibia (shin); the elephant is a sauntering graviportal mammal and its T/F ratio is only 0.60. In the race horse the lower leg bones have elongated considerably and the ratio has increased to 0.92. In really fast sprinters such as the gazelle the tibia is even longer than the femur and the stride approaches its maximum: here the ratio is 1.25.[26] The mountainous brontosaurs had a T/F ratio almost identical to modern elephants; the massive limbs and the structure of the limb joints in the two types of creatures widely separated by time were astonishingly similar and geared to supporting a crushing weight. The huge graviportal dinosaurs were also much straighter limbed than their more agile contemporaries, and less flexure could occur at the knee and elbow, which is also the case in the pillar-like columns of an elephant. This slowed the sauropod dinosaurs to an amble. In contrast, most of the hadrosaurs had ratios well over 0.80, and some of these bipeds equalled a race horse in their leg porportions, from which we are justified in concluding that they too were fast runners. Ostrich-mimics, as we might by now have anticipated, had a tibia exceeding the femur in length, and with a ratio of 1.12 the creature could probably have held its own in competition with the most rapid mammals. One particular *Ornithomimus*, with a shin:thigh ratio of 1.24 was a veritable gazelle of the dinosaurian world.[27]

Dale Russell's study of the skeleton and leg muscles of the Canadian Late

30. *Hypsilophodon*, a small, agile dinosaur, was a veritable gazelle of the dinosaurian world.

Cretaceous ostrich dinosaurs, and his functional comparison of struthiomimids and ostriches, has recently lent crucial support to the idea of fast-running ostrich mimics. Russell concluded that as far as speed was concerned these dinosaurs had rarely been equalled during the history of life; at least one member of the group, *Dromiceiomimus*, could probably run *faster* than the ostrich.[28] Other bipedal dinosaurs were undoubtedly very fast. The critical factor which caused Galton to bring *Hypsilophodon* out of the trees once and for all was the long shin region. Even for dinosaurs this by now has become a tell-tale characteristic of a running animal. With a shin:thigh ratio of 1.18 *Hypsilophodon* must have been able to build up a very fast speed. It was perhaps one of the only plant-eaters that could have out-paced the smaller long-legged coelurosaurs. *Hypsilophodon*, like the other dinosaurs, held its body horizontally as it dashed away with its tendon-stiffened tail raised in mid air. Like many runners, it sprinted on its toes rather than the soles of its feet.[29]

These figures are not absolute criteria for judging speed; it does not necessarily follow that identical ratios allowed their owners identical sprinting times in life, since a great many other factors have to be taken into account (not the least of which are the adjustments to be made when comparing two-legged and four-legged runners). Nevertheless, they give us a guideline. Most importantly, the use of such figures derived from the study of living and fossil mammals illustrates the way palaeontologists are now turning to mammals and birds and studying *their* adaptations to running in order to better understand the creatures which moved about the planet during the Age of 'Reptiles'.

4. The Dark Ages

Since we are mammals it is difficult not to adopt a chauvinistic attitude towards geological history and see the origin and subsequent evolution of our ancestors as an event of overriding importance. Man's present dominant station blinds him to the possibility that his predecessors were overshadowed by their reptilian and dinosaurian cousins. Mammals ultimately triumphed over reptiles, and we are loathe to admit that there was a time when mammals themselves were totally dominated, and not by inferior, but by physiologically superior creatures. It would be a mistake to imagine that dinosaurs had little to do with mammalian evolution. They were of the utmost importance in our history.

Mammals were in existence as early as the latest Triassic, 190 million years ago, yet for the first *one hundred and twenty million years* of their history, from the end of the Triassic to the late Cretaceous, they were a suppressed race, unable throughout that span of time to produce any carnivore larger than cat-size or herbivore larger than rat-size. Why were mammals *kept* persistently tiny, for the most part small shrew-like insectivores that were forced to venture out under the cover of darkness and to spend the daylight hours in the safety of burrows? Why was there no immediate radiation towards the array of forms we find today? Why were no 'elephants' living in the Jurassic, 'whales' in the Wealden, and 'sabre-tooth cats' in the late Cretaceous? Why did 'man' not land on the Moon in the Cretaceous? (These beasts would not have been replicas of today's versions, of course, but creatures fulfilling the same ecological role. Perhaps these archaic, counter-factual mammals are better styled 'elephantoids', 'humanoids', and so on.) Looking about us, we are made acutely aware of the mammal's potential for success. It has all the ingredients: a superb locomotor mechanism, an efficient physiology and intelligence to put any reptile to shame. Yet for the first two-thirds of the group's history practically nothing happened, the eventual take-off only following-on the demise of the dinosaurs at the end of the Cretaceous. Mammals played no part in the extinction of the dinosaurs. They had to wait passively in the wings until dinosaurs disappeared before they were able to exploit the same niche. It was as if a brake was suddenly removed, releasing mammals from their subservient position.

Our ancestors were relegated to an insignificant role in the Mesozoic. Dinosaurs were the masters of that world, creatures so efficient in physiology and locomotion that they snatched the world from the mammals' grasp and monopolised it for 120 million years. How they were able to do this, and thus set back our own development for so long, is only now becoming apparent as we begin to plumb the dinosaur's own origin.

The Elgin sandstone in Scotland is a classical fossil-collecting locality, where numerous Upper Triassic reptiles came to light in the last century. Many of these were aetosaurs, like *Aetosaurus* and *Stagonolepis*, reptilian 'pigs' encased in impregnable bony armour embellished with serrated protruding spikes, and with pig-like snouts for rooting through the soil. Although they were without question terrestrial root-grubbers that shuffled round on all fours, the hind legs were perceptibly longer than the front ones, a sure sign that their ancestors had been marsh-dwellers requiring powerful back legs for kicking and steering. Indeed, there were still aetosaur relatives living in the Late Triassic swamps, beasts resembling long-snouted crocodiles and known as phytosaurs. These forms comprise an important Triassic group, characterised by socketed teeth – for which reason Owen styled them thecodonts, to distinguish them from lizards, in which the teeth are fused to the jaw. From this 'reptiliferous' sandstone, as E. T. Newton at the Geological Survey expressively termed it, came another of these thecodont creatures in the 1890s, but one that differed in several crucial respects. Newton described this new Triassic reptile in 1894, calling it *Ornithosuchus*,[1] and notwithstanding the friable nature of the fossil bones, he was made quickly aware of this thecodont's deviant nature. *Ornithosuchus*, the 'bird crocodile', a name highlighting its intermediate nature, had retained some of the aetosaur's primitive features, but Newton observed many similarities to the carnivorous dinosaurs, particularly in the serrated teeth (*Ornithosuchus* was evidently a highly predatory reptile) and the skull, which he likened to a miniature megalosaur. Certain features were reminiscent of birds and crocodiles, although the skeleton looked remarkably like Huxley's little *Compsognathus* dinosaur.

Similarities between aetosaurs and the dinosaurs that followed them had always been recognised, and the new fossil bridged the gap admirably. The 'difficulty of separating the two groups', observed Newton, 'is increased by a study of this new Elgin reptile, which holds . . . a more intermediate position between the two series'. Although its skull and teeth were dinosaurian, *Ornithosuchus*' pelvic girdle and limbs were definitely thecodont in appearance. Here was one of the first clues to dinosaurian ancestry. Newton was inclined to make *Ornithosuchus* the most primitive dinosaur rather than the most advanced thecodont, even though he conceded that there was little in it. Newton's *Ornithosuchus* was very small, well under three feet, but an individual six feet long was found in another Elgin quarry in 1901.[2] Reappraising its primitive features, the Curator of Reptiles at the London Zoo returned it to the thecodonts, where it joined the aetosaurs and phytosaurs. The quarrel was really only one of personal preference, as Robert Broom saw in 1913:

> Though these two opinions seem at first sight to be at variance they are really pretty similar. Practically, it amounts to this, that in the Pseudosuchia [the group to which *Ornithosuchus* belongs] we have a group of primitive reptiles which, while they do not fit into any of the later specialised orders, have affinities with quite a number of other groups.

There cannot, I think, be the slightest doubt that the Pseudosuchia have close affinities with the Dinosaurs, or at least with the Theropoda [flesh-eaters].... In fact, there seems to me little doubt that the ancestral Dinosaur was a Pseudosuchian.[3]

Broom, the foremost South African palaeontologist, who specialised in Triassic reptiles and mammal-ancestors, was opting for the non-dinosaurian nature of *Ornithosuchus*. At the time, Broom was describing a still older pseudosuchian, which he named *Euparkeria*, occurring in the Lower Triassic in South Africa. *Ornithosuchus* and *Euparkeria* were obviously very closely related, and since Broom had much of the skeleton of his South African form he was able to build up a picture of the dinosaur-ancestor. Of the greatest consequence was the fact that *Euparkeria* had front limbs far shorter than rear ones. Broom observed that while the humerus and radius measured a total of 68mm, the femur was 58mm and the tibia 'apparently a little shorter than the femur and considerably more slender', that is, the hind limb was a little over 100mm overall. Broom was paving the way for the concept that the ancestors of the dinosaurs among the pseudosuchians *had already begun to walk erect by Early Triassic times*, 210 million years ago.

There was no argument about *Euparkeria*'s position, it was definitely a thecodont. Nevertheless, said Broom, 'there is nothing in the post-cranial skeleton that is not just what we should expect to find in the Dinosaur ancestor'. Broom assessed the beast's habits; '*Euparkeria* is in my opinion potentially bipedal, and was probably partly bipedal in its habits . . . the animal possibly ran on its hind feet.'[4] Broom saw this little thecodont chasing large insects such as locusts and grasping them in the air. '*Ornithosuchus*,' he concluded, 'was probably very similar in habit to *Euparkeria* and was even a little better adapted for fast running on its hind feet.' He thus set the precedent for believing that these agile bipeds were ancestral to dinosaurs, and concomitantly, that an erect posture was acquired before true dinosaurian status. *Euparkeria*-like thecodonts, he believed, were midway between true dinosaurs and the armoured aetosaurs, retaining very small armour plates on their back but losing most archaic thecodont features.

The versatile and unspecialised pseudosuchian thecodonts gave rise to other groups in Broom's imagined family tree. He suspected that it was from such fleet-footed thecodonts that the pterodactyls arose, by way of an intermediate gliding phase. He even suggested that this versatile thecodont group gave rise to birds, which retained the hopping gait of the dinosaur-ancestors. The versatile Pseudosuchia (literally 'false crocodiles') were also suspected of being crocodile ancestors. Although some of Broom's suggestions are now disputed (for example, that birds evolved so early in time), his views had a lasting influence during the next half century. His belief that *Euparkeria* and *Ornithosuchus* were ancestral to dinosaurs was readily accepted, along with his conviction that the dinosaur-ancestors were already erect. Colbert has summed up the current position:

In such thecodonts the hind limbs are elongated, and the forelimbs are relatively small. The body is pivoted at the hip joint, and the weight of the body in front of this pivotal joint was in life effectively counterbalanced by a long tail.

Compare in your mind's eye the gait of this early thecodont with the gait of a swift-running ground bird, like the road runner of our south-western states, and you can get a picture of early archosaurian locomotion.[5]

There can be little doubt that the Late Triassic *Ornithosuchus* was a bipedal animal. The construction of the pelvic girdle and the lateral head on the femur testify that the limbs moved in a fore-and-aft direction, rather than projecting horizontally at the sides of the body. Since the forelimbs were reduced, if the animal tried to walk on all fours the hind limbs would have outpaced the front ones, and the backbone would have been uncomfortably arched over the pelvis. In the last decade *Ornithosuchus* has again see-sawed back and forth between dinosaurs and thecodonts, but this only highlights the similarity between advanced thecodonts and early dinosaurs. Some would like to see it directly ancestral to *Tyrannosaurus* and *Gorgosaurus*,[6] whilst others consider it too specialised to be directly ancestral to dinosaurs.

The Early Triassic *Euparkeria* was a much more generalised beast, and it has been suggested that it gave rise, not only to the giant carnivores via *Ornithosuchus*, but also to the mountainous Jurassic herbivores like *Brontosaurus*.[7] Since the limbs of *Euparkeria* were proportioned very like those of *Ornithosuchus* (the humerus was 66% of femur length in *Ornithosuchus*, and a little over 65% in *Euparkeria*), there is every reason to suppose that it too would have experienced

31. In Early Triassic times, 210 million years ago, small pseudosuchian thecodonts became the first animals to walk erect. Although bringing the legs under the trunk increased the stride, it could only be used to the best advantage if a fast energy output could be maintained. *Euparkeria* (above) was simultaneously transforming both posture and metabolism. Related thecodonts were responsible for the explosion of the dinosaurs to the detriment of the mammal-like reptiles.

86

difficulty in walking on all fours. If it moved bipedally, it must have brought its legs in under the body. (These two features are interrelated, as Alfred Sherwood Romer clearly saw: 'The body could not, of course, be supported in a biped with the legs spread out at the side of the body.'[8]) It has been suggested that when it ran it tucked its legs into the sides of the body, but when it ambled the legs reverted to a sprawling position.[9] Certainly we should expect a transitional stage in the origin of erect bipeds, but whether we have been lucky enough to meet it in *Euparkeria* remains debatable.

Even granting that this small lower Triassic thecodont was not permanently bipedal, but adopted the ancestral sprawling gait when not hurrying, we still seem to have a beast in the process of becoming erect. Shifting the femur into the vertical plane to increase the stride was the first step to an upright pose. Obviously, we are witnessing a transition in the method of locomotion, as well as in the physiological make-up of the animal. Since an upright stance is correlated to greater energy expenditure, as Ostrom pointed out, we are forced to conclude that a physiological transformation was well under way by early Triassic times. The legs were being swung under the body to increase the stride while the metabolism was shifting towards sustained high energy production. *This would supply the power absolutely necessary if that stride was to be used to the greatest advantage, that is, for sustained fast running*. By the mid Trias, 200 million years ago, thecodonts had completed the process – they were warm-blooded, fast-metabolising beasts.

Just why did these dinosaur-ancestors become bipedal 200 million years ago? They had apparently acquired their longer hind legs for quite another purpose. Very primitive thecodonts were swamp-dwellers.[10] Like crocodiles, they had developed stronger hind limbs for steering and kicking in the water and a strong muscular tail for propulsion.[11] These marsh-infesters shared their world with forms totally unfamiliar to us today. Giant labyrinthodont amphibians lurked in these swamps, creatures that resembled severely flattened crocodiles. Many were so flat that they were unable to lower their jaws – they were already resting on the ground. Instead, the skull had to be raised if the animal was to eat![12] In the shallows of these Triassic marshes browsed miniature reptilian hippopotamuses' with walrus-like tusks. These corpulent vegetable-feeders were obviously good to eat because the early aquatic thecodonts are usually found associated with them as fossils. But as the Triassic became warmer, these swamps slowly dried up; the amphibians were already scarce by the middle of the period and the staple diet of the carnivorous thecodonts, the 'hippos', became extinct. Drastic measures were required if these thecodonts were to survive; they ventured out on to land for the first time in search of new prey, and this they met in abundance.[13] On land they suddenly found themselves possessed of one tremendous advantage inherited from their aquatic days – long hind legs and a powerful tail. They were already partially adapted to running on their hind legs, and this they would do, initially for short bursts to catch prey or when danger threatened.[14] As an incipient biped, the small *Euparkeria* found its lifestyle transformed and, moving into the dry upland areas, it was able to put its striding

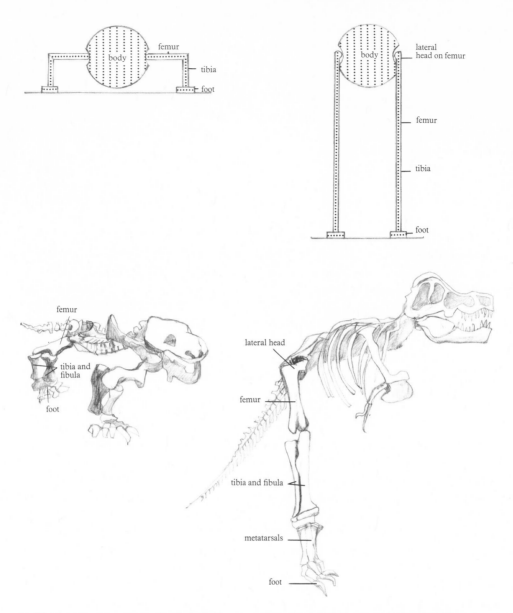

32. Dinosaurs had departed radically in their posture from the archaic Permian reptiles. The Karroo pareiasaur – a cumbersome herbivorous reptile – held the humerus and femur horizontal (left). *Tyrannosaurus* stood erect on vertical limbs; this required the development of a right-angled head on the femur.

limbs to good use on the firmer ground. This was obviously a great success. Already by the mid Trias there were fast sprinting 'rabbit' thecodonts living in Argentina; small reptiles with exceptionally long hind legs, even more stilt-like than in ostrich dinosaurs.[15] There can be little doubt that these dinosaur-forerunners had already acquired the high energy producing metabolism so essential to an erect sprinting creature. They had broken the thermal barrier. These thecodonts could outstride any contemporary and run their choice of prey

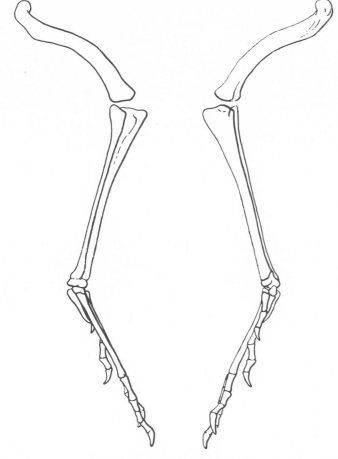

33. The recently discovered hind leg of the otherwise unknown 'rabbit' thecodont *Lagerpeton*. The in-turned femoral head and long shin and foot are incontestible evidence that fully erect, fast-running thecodonts were living in Argentina by Middle Triassic times. The thermal barrier had already been broken.

to the ground. And with legs held under the body rather than at the sides the beasts could grow to gigantic proportions; standing upright for the first time was the key to unprecedented growth. This act was one of the most momentous key innovations in vertebrate history, the repercussions were to be tremendous. It allowed the thecodonts and their ancestors to conquer a world.

At some point between the emergence of the primitive thecodonts from the water and their evolution into rapid sprinters the dinosaurs made their debut, emerging as successive waves from the ranks of the thecodonts.[16] This area of study is currently in a state of flux. Many of the distinct dinosaurian types can be traced with little change back to the mid Trias. This has suggested to some experts that they had separate thecodont ancestries. If multiple origins are the case, however, the respective ancestors were probably closely related. It has lately been suggested, on the grounds of the many features – particularly relating to the ankle joint – common to *all* dinosaurs, that the group had but a single ancestor.[17]

It first struck Harry Govier Seeley, Professor of Physical Geography at Bedford College, London University in 1887 that dinosaurs were really of two distinct types; the Saurischia (flesh-eaters and brontosaurs) and Ornithischia (iguanodonts and hadrosaurs, ceratopsians, ankylosaurs, and stegosaurs), these two major types differing in their pelvic girdle.[18] Thus it appeared that the term 'dinosaur' was unnatural, an artificial grouping of two sorts of unrelated giants. In only the last few years the question has become vastly more complex – and theories now run the entire gamut from a single ancestor to many different ancestral groups for dinosaurs. The important criterion for deciding whether the term 'dinosaur' has any validity is the relationship the various ancestors bear to one another (if multiple origins are conceded). Multiple origins are no bar against grouping the respective descendants together; in fact, single origins are the exception rather than the rule in the appearance of major groups of animals. If the descendants all come from a restricted ancestral group – even if they evolve from different members of it – we are justified in grouping them together. Since all dinosaurs were probably fully erect endotherms, it is simpler to suppose that these key innovations existed as a potential or were already present in a restricted ancestral group within the thecodonts, and that successive waves of dinosaurs carried them forward into the major lineages. If this is conceded, the term 'dinosaur' is a natural one.

The potential inherent in an upright stance, mated as it was with a new physiological attainment, resulted in an explosion of diverse forms that rapidly filled and monopolised every conceivable type of terrestrial niche open to large, active animals. The little *Euparkeria*-like reptiles entered a world where their descendants could achieve the speed of struthiomimids, the size of brontosaurs, and the ferocity of tyrannosaurs. The consequences of the migration of waves of small and medium sized active thecodonts and later dinosaurs away from the low-lying wet regions and into the dry uplands were to prove catastrophic to the reptiles already occupying these regions, not the least of which were the still semi-sprawling mammal-ancestors. The small thecodonts and dinosaurs were moving into their terrain, and the mammal-ancestors found themselves no match for these nimble, active endotherms.

The therapsid reptiles, among them the ancestors of mammals, underwent a Golden Age in late Permian and very early Triassic times. For thirty million years preceding the advent of the thecodonts and dinosaurs, it was these therapsid or mammal-like reptiles that totally dominated the earth. If we turn to the South African Karoo system, a series of beds laid down as sediments filling a subsiding basin, we find a nearly complete record of life from the Permian through to the late Triassic. The sequence of beds, housing their respective faunas, is so complete that we can actually witness the vying for dominance by various animals over the critical period, lasting 50 million years. The most striking feature of life in the Karoo basin in the later Permian is the prevalence of therapsids, practically to the exclusion of all else. This one order flourished in a profusion of forms, having radiated into all niches, from those of huge archaic herbivores to small grubbing insectivores. The most prominent herbivores were huge dicynodonts with walrus-like tusks, a group that itself radiated into a variety of types, from capacious plains-dwellers to equally stocky aquatic reed-feeders. The Karoo was simply teeming with these two-tuskers in late Permian and lower Triassic times. In places their bizarre tusked and beaked skulls litter the landscape, suggesting that these corpulent beasts roamed the Karoo in substantial herds. In some of the beds dicynodonts account for eighty percent of all skeletons found.

Such herds of game naturally attracted carnivores, themselves distantly related therapsids. Many of these lightly built predators resembled semi-sprawling wolves. With more gracile bodies and slender limbs, these carnivores obviously took a heavy toll of the herds of archaic game reptiles. In many predators the upper 'canines' had become elongated, sabre-tooth fashion, and obviously functioned as daggers in penetrating the thick hide on the dicynodonts. Few carnivores were small, the vast majority were dog-sized or larger and well able to prey on the cumbersome plant-croppers. Later in the Permian and especially in the Triassic these carnivores became more prevalent and diverse in appearance, and many types began experimenting with novel approaches at increasing efficiency.

There is at least circumstantial evidence that these carnivores were attempting to increase internally generated heat. In the skulls of these mammal-like reptiles, the cheek bones begin to grow across the roof of the mouth to separate the mouth from the nasal passage. The advantage of a secondary palate is manifold. The

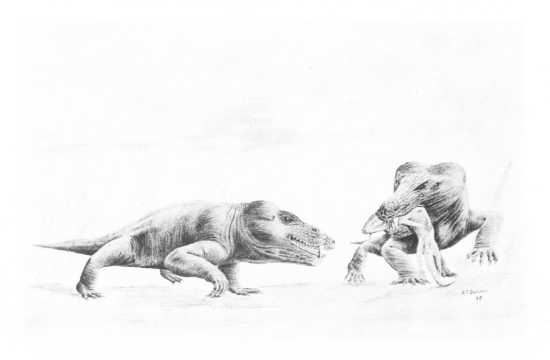

34. *Cynidiognathus*, thriving in Early Triassic times, was a semi-sprawling wolf-like therapsid about 4 feet in length. Therapsids began to wane as the agile thecodonts moved into their terrain.

animal no longer had to cease breathing in order to eat. With a secondary palate the food could remain in the mouth for longer, while breathing continued unimpaired. This allowed the food to be chewed before being swallowed. The teeth in these skulls displayed the first signs of increased complexity, with the cheek teeth developing additional cusps. This suite of characters suggests that uninterrupted breathing and well-chewed food were becoming necessary; the animal was stepping up its energy output.[19] At what stage – if at all – these mammal-ancestors grew hair remains conjectural, although we are aided by a single observation. In the early Triassic, there lived in the Karoo at least two tiny therapsids that exhibited one additional feature. In the skull, the cheek bone was perforated externally by many forwardly-directed openings. D. M. S. Watson, who first drew attention to these therapsids in 1931, speculated on the function of the perforations. They carried, he ventured, a series of very large blood vessels and nerves to the snout. There must have been a considerable extension of the muzzle beyond the bony skull, and this shrew-like snout was probably equipped with muscles and sensitive whiskers.[20] If – and the tortuous train of reasoning suggests that this must remain highly tentative – whiskers grew on the snout, hair undoubtedly covered the body: whiskers are only modified hairs. This core of circumstantial evidence intimates that therapsids were experimenting with an endothermic physiology, although it may still have been imperfect at this stage. Living monotremes, like the egg-laying duck-billed platypus and spiny anteater, are thought to have descended from therapsids independently of other mammals, and to have inherited their physiological make-up. Since monotremes lack a cooling system, being unable to sweat or pant, it is not unreasonable to assume that the same was true of the Triassic mammal-like reptiles.[21]

The Permian Karoo suffered periods of intense cold, so this incipient attempt to step-up the body's heat output and stop heat loss through the skin, however inefficiently, conferred on the therapsids a distinct advantage over the lizard-like reptiles.[22] Yet even advanced therapsids remained at most semi-sprawlers,[23] while the Australian platypus retains a completely reptilian sprawl. It seems that in the late Permian therapsids were more concerned with raising their temperature to offset the cold than with sending increased power to the limbs.

It was into this therapsid-dominated Karoo world that the thecodonts burst. During the Trias the climate ameliorated and the temperature became less prone to drastic fluctuation, so the large thecodonts and dinosaurs found no need to insulate their skin as the mammals had done. The thecodonts, moreover, had rapidly adopted an erect posture and had stepped up *their* energy output specifically to power the long striding limbs, resulting in a marked increase in speed and agility. (Since an increase in power demanded an endothermic physiology for optimum biochemical functioning, this was simultaneously acquired.) As waves of thecodonts passed into therapsid terrain, there was a period of intense competition.[24] The dinosaur-ancestors were much better equipped, and their diversification was accompanied by an equal decline in therapsids; by the mid Trias there were few left, at the end of the period they were extinct. The Golden Age was brought abruptly to an end: the mammalian Dark Ages were beginning.

By the mid Trias the thecodonts found themselves in competition with their own descendents, the dinosaurs, and had to bow out in their turn. This succession was not a violent affair. Thecodonts ousted therapsids from their niches and moved in themselves, rather as the hardier grey squirrel has usurped the red squirrel's niche in Britain. Niche 'stealing' is a common occurrence in faunal succession, where creatures better adapted to the prevailing environment topple their less-endowed competitors in bloodless contests. Herds of dicynodonts made way for herds of prosauropod dinosaurs. Already by the end of the mid Triassic *Melanorosaurus* – a 40 foot forerunner of the brontosaur – was a conspicuous feature on the South African landscape. Similarly, the 'wolf' therapsids were replaced by carnivorous thecodonts and dinosaurian flesh-eaters in turn. It was a race to an upright posture, with the winners inheriting a world. Our ancestors lost.

The dinosaurs conquered both upland and lowland. They exploited every niche that was open to medium and large-sized creatures. Physiologically they showed a degree of sophistication that was lacking in therapsids and early mammals. Unlike the latter, dinosaurs undoubtedly did have the ability to unload excess heat after exercise, perhaps – like birds – by means of an air sac system permeating the body. This permitted them a level of activity almost unheard-of in a land animal up till then.

Meanwhile, the newly evolved mammals were plunged into their Dark Ages. The one niche dinosaurs could not exploit was that of mouse-sized creatures; below a critical size dinosaurs would have lost too much heat through their naked skin. Here was the only opportunity open to mammals for survival. They retreated into this niche. Thus the first true mammals, which appeared on the earth in Late Triassic times (descendants of at least two lines of therapsids passed into the mammalian grade), were very unwolf-like. Throughout the 120 million years of the Dark Ages mammals were restricted to a very small size and lived in dry upland environments. The inhabitants of such environments are always ill-represented in the fossil record; it is the low-lying estuaries and coastal regions that accumulate rock-forming sediments. Consequently, early mammal remains are exceedingly rare, for the most part just resilient teeth. 'Until recently,' estimated A. W. Crompton, now head of Harvard University's Museum of Comparative Zoology, 'the entire collection of mammalian fossils from the "Dark Ages" could be housed in one medium-sized shoe-box.' This is a reflection, not only of the paucity of numbers, but also of the miniscule size of the oldest mammals' teeth.

One region where very early mammal teeth have been located is Glamorgan in South Wales. As the Triassic gave way to the Jurassic, the great Tethys Ocean – extending from the Caribbean to Russia – began to creep northwards, drowning much of Southern Britain and driving the mammals north with its advance. At least two types of tiny mammal became stranded on a chain of small islands where South Wales now stands. These island homes were made of limestone riddled

35. The elongate jaw of *Morganucodon*, one of the first mammals. The jaw, 15 mm long, comes from the late Triassic Welsh fissures.

with crevices, and as the mammals perished their remains were washed by rainwater into these crevices.[25] Workmen quarrying the limestone for roadmetal expose the fissures and bring to light tons of tooth-bearing rock. The teeth, from beasts named *Morganucodon* and *Kuehneotherium*, are little larger than 1mm. Many samples of the Welsh fissure material have been processed at University College, London since its significance became appreciated twenty-five years ago.[26] Vats of acid are used to dissolve many pounds of rock to release the acid-resistant teeth (and bone fragments), which are then sorted and graded. The teeth have complex cusp patterns and the upper molars shear precisely past the lowers, a fit that allows efficient mastication of the food. Well-chewed food is important in mammals, because it permits speedier energy release.

Even from the teeth alone we know that we are dealing with mammals and not therapsids. In reptiles, including therapsids, the teeth are continually replaced along the jaw throughout life, because the reptile grows continuously throughout life. Each succession of teeth must be larger than the last to keep pace with the lengthening jaw. This system, however, does not permit a close fit of the upper and lower teeth, since they are all in various stages of eruption. In mammals, the teeth are only replaced once; the second set remain until the animal's death as a permanent feature and fit together perfectly. This is possible because mammals rapidly reach their maximum size. Since the young mammal (which is born with a large head anyway) is nourished by maternal milk, it can delay the eruption of its permanent teeth until almost fully grown. In *Morganucodon* and *Kuehneotherium* from the Welsh fissures we have evidence of milk and permanent molars. Since milk molars imply parental care, we are, by definition, dealing with mammals.[27] But, as with all transitions, the process must have been a gradual

36. *Morganucodon*, stranded on its Welsh island in the Triassic Tethys Ocean, probably scavenged or ate beetles.

one, and these oldest mammals exhibit a vestige of their reptilian inheritance. Later in the individual's life, the 'permanent' premolars and sometimes molars as well begin to drop out, as if paving the way for another set.[28] But these never appear, leaving the tiny mammal with a gap in its mouth. We have caught these first mammals in the very act of suppressing the last of their reptilian replacements.

Articulated skeletons are conspicuous by their absence in these Welsh fissures. However plentiful the earliest mammals' teeth, complete skulls remain elusive. But in another corner of Crompton's 'shoe-box', under a welter of teeth, *can* be found three very precious skulls. The first was found by Father Oehler of the Catholic University in Peking while searching for mammal-like reptiles in Upper Triassic beds in the Yunnan Province of China in 1948. Father Harold Rigney, the Rector of the University and organiser of the expedition, arranged for the skull to be smuggled to the West in 1949 prior to his imprisonment by the communists after Peking fell. This skull was equipped with teeth like those in the Welsh fissures, thus the small beast became known as *Morganucodon oehleri*.[29] The other two skulls, of similar age, are from the South African Red Beds. Discovered in 1962 and 1966, *Megazostrodon* and *Erythrotherium* are as tiny as the Welsh mammals, with skulls no more than 20mm long. It is surprising that they were ever spotted at all. In fact, it was only by a stroke of luck that the first of the African skulls was recognised; it happened to be embedded in a slab of rock which was taken to the South African Museum because it contained an ornithischian dinosaur.[30]

Although fragments exist, there is only one partial skull known from rocks representing the next 100 million years; it is only in the late Cretaceous Mongolian beds that skulls are again found. Times were indeed dark for mammals. Unable to radiate into large-body niches because these were already monopolised by dinosaurs, mammals were forced to bide their time. It was not spent unproductively, however. During this period they perfected their endothermic physiology by developing an efficient cooling mechanism, involving panting and sweating, and slowly gained an upright stance like the dominant dinosaurs. But all the while they remained an insignificant part ofthe Mesozoic fauna. Being so small, they were probably forced out at night to escape being killed and eaten; there were plenty of larger lizard-like creatures at this time that hunted by day but became sluggish in the cooler night air. Our New Zealand Tuatara 'lizard' is a relict of an ancient group that flourished in the Mesozoic. Although reduced to a single species now, these reptiles were once extremely common. Many dinosaurs were simply too large to bother about such small fry, but the smaller coelurosaurs like *Compsognathus* had it in their power to catch mammals. It was not until the Cretaceous, however, that we find signs that mammals were hounded even into the night. They were terrorised, moreover, by creatures *more* cunning than themselves.

Ostrom's tenacious *Deinonychus* was ancestral to a variety of small but no less aggressive carnivorous dinosaurs in the later Cretaceous, including the Mongolian *Velociraptor* found entangled with its victim. *Velociraptor*'s quarry was obviously small herbivorous dinosaurs, and Ostrom suggests that *Deinonychus* itself had a preference for small plant-eaters. There are reasonable grounds for supposing, however, that some of *Deinonychus*' other descendants preyed upon mammals.

It has become a cliché that dinosaurs were so small-brained that they could have been only dimly aware. But the degree of sophistication found in *Deinonychus* – its agility, speed, and one-legged balancing ability – seems totally inconsistent with the popular conception of dinosaurian mentality. The nervous coordination necessary for such stunts demanded a far more complex brain than that possessed by the majority of dinosaurs. The intricate manipulation of the fingers alone argues strongly for a well-developed centre of coordination, in addition to huge eyes to oversee the operation. The braincase of *Deinonychus* itself remains unknown, so we have to rely on some of its equally talented relatives for knowledge of this vital factor. One of the latest Cretaceous descendants of *Deinonychus* was a little known creature called *Dromaeosaurus*, the 'emu reptile', named after its supposed avian counterpart, and a beast that has now lent its name to the entire group – the dromaeosaurids. Its bones were originally found by Barnum Brown during his foray into the Red Deer River locality of Alberta in 1914, but they were only fully described by Edwin H. Colbert and Dale A. Russell in 1969. All Brown collected of this carnivore were some foot bones and a partial skull, which fortunately had a large part of the braincase intact. The striking feature of the braincase was its shape, which was quite atypical among dinosaurian flesh-eaters. The bony cavity that housed the brain was broad, especially in its posterior or medullary region. 'If the posterior moiety of the brain even approximately filled this space,' exclaimed Colbert and Russell, obviously caught unawares by this most undinosaurian shape, 'it must have been very broad indeed.'[31] This is the first inkling that there were dinosaurs with brains larger than the proverbial walnut. Like *Deinonychus*, *Dromaeosaurus* was a skilful killer, with trenchant teeth and a strongly recurved hind claw for lacerating, so it is interesting to correlate its undoubted agility and dexterity with the increased mental capacity. Dromaeosaurids were obviously a force to be reckoned with in later Cretaceous times; nearly as fast as struthiomimids, relatively more formidable than any flesh-eater, and with an ability to scheme unheard-of in their ancestors or contemporaries, dromaeosaurids were more frightening in their way than even the towering tyrannosaurs. And they were perhaps more common than their rare remains would have us believe. It was their environmental preference that counted against them being preserved as fossils.

Interest in dromaeosaurids began to be felt in the late 1960s, when it became apparent that there was a whole new group of dinosaur fossils lying relatively untapped. So in 1968 Dale Russell led a team from the National Museum of Canada into Alberta specifically to seek out these rare creatures. For six weeks his crew collected nothing but isolated bones, then in the last few days in the field they had the good fortune to locate the scattered skeleton of *Stenonychosaurus*, another of *Deinonychus*' little known descendants.[32] This small predator was only five or six feet long, but like other dromaeosaurids had a sizable head, perhaps nine inches long. The skull was distinguished by disproportionately large eye sockets which, if the eyeballs had filled them completely (and there is every reason to suppose that they did), would have given the little dinosaur eyes two inches wide. The eye was comparable to an ostrich's, which is the largest of any land animal today. Since the distance between the eyes was abnormally wide and the snout correspondingly triangular, the eyes were directed forward with their visual fields overlapping. This resulted in binocular vision. Eyesight was evidently acute and the ability to judge nearby distances accurately was used in conjunction with the opposable fingers. Sharp vision and this novel stereo modification were essential if the dinosaur was to catch and manipulate prey in its hands. This extraordinarily advanced state of affairs obviously required a well-developed central control, so it comes as little surprise that the wide interorbital region of the skull housed a greatly enlarged brain, with optic lobes expanded ventrolaterally (as in birds) and relatively huge cerebral hemispheres dorsolaterally. Until recently this was the maximum development of the brain known among dinosaurs. We must not lose sight of the fact that the vast majority of dinosaurs were endowed with negligible aptitude. This is staggeringly true in most cases. The hadrosaur had a brain weighing only $\frac{1}{20\,000}$ of its body weight; for the colossal brontosaur the proportion was $\frac{1}{100\,000}$.[33] With figures such as these it is easily understood how it came to be accepted that the dinosaur was an 'unthinking', almost totally automated creature, incapable of the sensitive co-ordination and cold-calculation found in a mammal and bird. 'The brain', stated Romer, referring specifically to brontosaurs, 'is small in all reptiles but excessively small in these dinosaurs in proportion to their size. Very likely the brain did little except work the jaws, receive impressions from the sense organs, and pass the news along down the spinal cord to the hip region, from which came the nerves working the hind legs.'[34] The dromaeosaurids were another matter entirely. Having risen to an avian level of intelligence, they had left the other dinosaurs way behind. A cast of the *Stenonychosaurus* brain, pieced together from braincase fragments, suggests that this dromaeosaurid had a brain volume of 49 cc and, consequently, that the brain itself weighed as much as 45 grams in life. Since emus the size of *Stenonychosaurus* are unlikely to weigh more than 45 kg (and this is an overestimate), the dromaeosaurid brain may have weighed more than $\frac{1}{1000}$ of its body weight. The large-eyed emu has a brain weighing about $\frac{1}{1227}$ of its body weight, which suggests that the emu would come in second to *Stenonychosaurus*, if not in the intelligence stakes, then at least in brain size.[35] A dinosaur with a brain of this size certainly flies in the face of tradition. Even recognising that smaller creatures have proportionately larger brains, still the dromaeosaurid was considered unparalleled in the dinosaurian world, having a brain 100 times heavier, relatively speaking, than the brontosaur. Moreover, particular areas of the dromaeosaurid brain, notably the cerebral hemispheres, were enlarged for controlling the sophisticated behaviour and governing intricate motor coordination. (In 1972 Russell proceeded to demonstrate that ostrich dinosaur brains were still larger, the cranial capacity of *Dromiceiomimus* exceeding an ostrich's.[36]) These Mesozoic bird-mimics and dromaeosaurs certainly possessed an intelligence compatible with an advanced social organisation. Quite probably, herbivores like *Struthiomimus* were flocking dinosaurs ('herding' hardly seems appropriate for such delicate bird-like creatures), while both carnivores and herbivores were capable of parental care, and a structured family grouping.

Stenonychosaurus was probably only one of many such predators, but of these creatures that shared its world we know very little, with the exception of one Mongolian form. *Saurornithoides* (the 'bird-like reptile', a name betraying its appearance), found by an American Museum expedition to Mongolia in 1922, was a close relative of *Stenonychosaurus*, equally large-brained and with the characteristic wide eyes and grasping hands.[37] As with so many of these dinosaurs, its significance was not appreciated until the late 1960s. Russell argues that both these dinosaurs – having manipulating fingers, stereo vision, and a sound intelligence – were capable of catching and eating mammals. It had previously been assumed that mammals eluded their enemies by venturing out at night, when the temperature dropped and reptiles became sluggish. But since dinosaurs probably were warm-blooded, they too could hunt by night. Perhaps these smaller dromaeosaurids slept by day like many living mammals and only stirred at dawn and dusk. Their large eyes (a good but not incontrovertible sign of nocturnal habits) would be used to pick out scampering mammals in the half

37. The Mongolian *Saurornithoides*, with its avian intelligence, stereo vision and manipulative fingers, may have hunted nocturnal mammals.

light. The grasping hand was probably used to catch the mammal and pass it to the mouth, or to pin the victim to the ground so the jaws could snap it up.

Seen in a broader perspective, active predation, however, played a secondary role in the suppression of the mammals through the Dark Ages. Dinosaurian control rested more in the nature of passive subjugation; they held the niches for large-bodied animals and barred the radiation of the mammals. This was still the situation in the late Cretaceous, when a mysterious calamity swept away the one obstacle to mammalian take-off. Something exterminated the dinosaurian race.

5. The stranding of the Titans

In July 1877 Othniel Charles Marsh, the Professor of Palaeontology in the newly opened Peabody Museum at Yale University, published a 'Notice of a New and Gigantic Dinosaur' in the pages of the *American Journal of Science*. This modest paper, little more than one side in length and written in Marsh's usual unadorned prose, began the most productive decade that American palaeontology has probably ever seen. 'The Museum of Yale College,' Marsh wrote, 'has recently received from the Cretaceous deposits of Colorado a collection of reptilian remains of much interest. Among these specimens are portions of an enormous Dinosaur, which surpassed in magnitude any land animal hitherto discovered.'[1] Marsh had only a few vertebrae and other fragments but their size suggested an animal of gigantic proportions. Each vertebra was a foot in length and was flanked by laterally projecting processes spanning three feet. Marsh guessed fifty or sixty feet for the entire creature and christened it appropriately *Titanosaurus montanus*, the 'titanic reptile'.

Marsh's paper stands as a landmark not because of its revelations (there were few apart from the reptile's size) but because it marked the beginning of an era. The New World was about to experience the discovery of giant four-footed dinosaur remains in such profusion that all previous finds would be completely eclipsed.

Nor was Marsh even the first to realise the existence of these Mesozoic giants. At the time of his writing gigantic reptiles had long been known in England, although it had only recently become appreciated that the largest of them were, in fact, dinosaurs like the *Iguanodon* and *Megalosaurus*. Before Richard Owen gave his classic account of the characters of the dinosaur at the British Association meeting of 1841, he told of his attempt to sort out the jumble of large vertebrae characterising almost every collection. After removing those of the marine saurians and his three species of dinosaur, Owen was left with what appeared to be huge crocodile vertebrae. One crocodile vertebra in particular impressed him because it was hollowed out, 'cancellated' or 'pseudo-pneumatic' as he called it, a feature completely unknown in a large saurian. He had first come across such a vertebra in a private collection belonging to a Mr Saull. The fossils had been weathered out of beds of Wealden or Early Cretaceous age by the sea and then rolled back and forth by the waves until the smooth bones were washed up on the Isle of Wight beach. These hollowed vertebrae led Owen to distinguish yet another species of giant saurian from his new-found dinosaurs, thereby adding one more enormous creature to his spectacular zoo of resuscitated monsters. Owen thought that these cavernous vertebrae were the remains of archaic crocodiles, although his predecessors had arrived at quite different conclusions. In the mid 1830s Dean Buckland had in his possession an arm bone and gigantic rib of one of these creatures; like the *Megalosaurus* bones they had been found near Woodstock in Oxford and had become part of Buckland's collection in the Oxford University Museum. On seeing these expansive bones Baron Cuvier unhesitatingly declared that they were the remnants of a prehistoric whale, a diagnosis that is somewhat surprising.[2] The strong similarity between these inordinately long ribs and those of a whale must have totally overridden the objection that a creature of such modern appearance could never have existed as far back in time as the distant Mesozoic. Owen, too, noticed the similarity to whale ribs and also that the texture of the bone, covered with a compact outer crust, was a cetacean feature. Cuvier's prestige was second to none and his opinions, even after his death (in 1832), had a persuasive force. Nevertheless, Owen possessed the vertebrae and they looked nothing like a whale's, so he was able to remove this rather obvious anachronism from the ancient world. But although the whale was ousted the supposed cetacean features persisted in the Mesozoic fossil. The vertebrae, observed Owen, equalled those of a full-grown whale in size and seem 'to indicate that the present gigantic marine Saurian must have had a capacious and bulky trunk, but propelled by a longer and more Crocodilian tail than in the modern whales'.[3] The similarity was acknowledged by Owen in the name he chose for the creature: *Cetiosaurus* or 'whale-lizard'. Owen supposed these reptilian colossi were strictly aquatic, living a whale's life in the open sea. They swam, he thought, by means of a large vertical tail fin at the end of a streamlined and naked body. We know that he supposed that the ichthyosaurs possessed this sort of fin because Waterhouse Hawkins' surviving ichthyosaurs at Crystal Palace have long whip-like tails with a vertical fluke. Owen envisaged the ichthyosaurs as shallow water dwellers, occasionally crawling on to the shore to bask in the sun and this is the pose that has been adopted at Crystal Palace, with the lake lapping around the ichthyosaurs' flippers. The long tail terminated by a fin is undoubtedly how Owen imagined the cetiosaur powered. Its webbed feet were presumed to be steering organs as they are in the marine iguana of the Galapagos Islands (not that he found any evidence of webbing in the cetiosaur, although he did have some limb bones at his disposal). In Owen's eyes the whale-lizards were better adapted to the sea than even the dolphin-shaped ichthyosaurs, and at this stage he speculated that they were carnivores placed in the prehistoric seas specifically to keep these lesser saurians in check.[4]

But Owen was wrong, as he soon found out. Baron Cuvier's renowned method of reconstructing the whole animal from a single bone failed to work in this instance simply because Owen had no precedent to guide him. Owen relied on crocodiles, lizards and whales for his models, so his cetiosaurs, known only from a few vertebrae and limb bones, inevitably became very hybrid in appearance. It is easy enough to reconstruct a fossil bird from a single bone when the avian blueprint is already to hand. (This Owen did to great acclaim. From a *six inch splint of bone with broken extremities* Owen rebuilt the extinct running bird

Dinornis in 1839; three years later a complete skeleton, 11 feet tall, was excavated in New Zealand, completely vindicating the Professor's published restoration and belief in the former existence of huge ground birds on that island. The popular press elevated Owen to the scientific pantheon for this single act of scientific sleight of hand.) Alas, when he attempted the same stunt with the unparalleled Mesozoic dragons he had no blueprint and very few guide lines and was destined to go hopelessly astray.

Owen's model of the whale lizard came under attack as further evidence suggested more realistic reconstructions and different affinities. New limb bones were found in Oxford in 1848 which finally forced him to abandon his 'cetacean hypothesis' of a marine, whale-sized crocodile, so that when he began his monumental six-part *British Fossil Reptiles* in 1849 (a series that was to reach completion only in 1884) he failed to allude to the whale analogy at all. But he was still convinced that the cetiosaur was a crocodile, although a crocodile with a difference. The size of the new limb bones suggested that it looked something like a dinosaur and he began, tentatively at first, to shift the cetiosaur in this direction. In Wealden times, he wrote, when the *Iguanodon* was abroad the land, there existed 'a Saurian reptile of dimensions at least equalling those of the *Iguanodon*, but with modifications of the vertebral column, from the middle of the back to the tail, departing from the Dinosaurian and approaching to the Crocodilian type'.[5] In what sense Owen understood transitional creatures is not easy to grasp from our post-Darwinian vantage point. Cuvier totally opposed the notion that the remarkable similarity in all vertebrate skulls, regardless of whether fish, reptile or man, resulted from any 'higher law of uniformity of type', as Owen called it, believing instead that bones serving similar functions were convergently similar. Owen disagreed. The pattern of skull bones is so similar in all vertebrates that some general plan was discernible. All vertebrate skulls are variations of an Ideal Archetype that the Divine Architect used as a blueprint at Creation.[6] Owen was more concerned with the way these homologous bones differed from one another than the mechanisms bringing about these changes. Nevertheless, he was prepared to go further than Cuvier and admit that natural forces were at work moulding the basic archetype into particular shapes. Therefore a creature such as *Cetiosaurus* midway between crocodiles and dinosaurs displayed some crocodilian and some dinosaurian divergences from the common plan, although it is a moot point whether Owen at this time (1846) would have conceded that *Cetiosaurus* was an evolutionary half-way house. (Despite his public hostility to the Darwinian cause, Owen often lapsed into evolutionary thinking in his later writings on fossil reptiles. It seems that it was the Darwinian *mechanism* that he objected to, based as it was on randomness, more than the *fact* of evolution.)

The same beds that produced Buckland's original 'whale' yielded another skeleton in 1868. Profesor John Phillips, Buckland's successor at the Oxford Museum, found a thigh bone as tall as a man in the Jurassic rocks at Enslow, eight miles north of Oxford. At five feet four inches long and one foot in diameter, this limb bone was unrivalled in the entire saurian world and its discovery proved

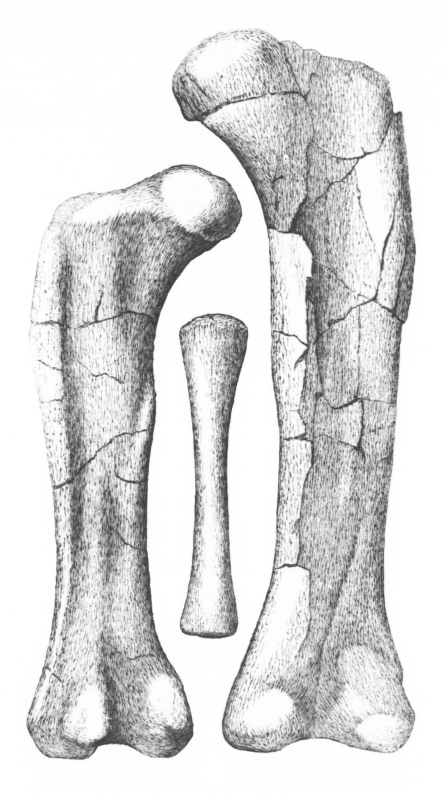

38. John Phillips' cetiosaur thigh bones, found near Oxford in 1868. The largest femur was as tall as a man.

more than adequate incentive to continue working the quarry. For over a year the rest of the creature eluded the searchers but Phillips was eventually rewarded by the discovery of a large part of the creature's skeleton. Since the days of the Crystal Palace festivities, the great extinct lizards had fired the public's imagination and new discoveries were eagerly reported in the popular press as well as scholarly journals. John Phillips' new cetiosaur naturally aroused considerable interest. Writing in 1870 in the *Athenaeum*, a peculiarly Victorian literary and scientific news journal, Phillips vividly recalled the quarry in which 'the largest animal that ever walked upon the earth' had reposed.[7] Scattered over the ground were bones five feet in length, vertebrae almost a foot in diameter, and 'monstrous ribs' five feet long even though broken at both ends (it is easy to see how the confusion arose in Cuvier's mind with such whale-sized ribs). All in all, said Phillips, it 'seems like the burial place of the great father of lizards.' Like other cetiosaur finds, there was no accompanying head, nor even any teeth found near the skeleton which could establish once and for all the affinities of the animal. However, Phillips possessed a good collection of stout limb bones testifying to the creature's four-footed pose, and as a terrestrial creature with massive limbs it became a prime candidate for Owen's Dinosauria. But Owen could not be absolutely sure that he had a new dinosaur until he knew how many fused vertebrae were present in the hip region.

It was the discovery of a new skeleton, not of *Cetiosaurus* itself, but of a supposed cetiosaurid relative, that forced Owen's hand. In 1874 the Swindon Brick and Tile Company, which worked the Kimmeridgian Clay in Wiltshire, wrote to Owen relating the accidental discovery of saurian remains in one of their clay pits. The bones were encountered ten feet below the surface and the quarryhands, realising the fragility of the fossils, had covered them up again to stop the bones crumbling in the air, whilst requesting Owen to come at once. In the ensuing months the British Museum acquired several tons of these bone-bearing rocks. Easing the bones out of the rock was no easy task and a 'mason-sculptor' was kept fully occupied for a year at the task. Eventually the bones were prized out and Owen at last saw his cetiosaur's hip region. The tail vertebrae were in perfect condition and showed the presence of a powerful whip-like appendage in life. The tail, he explained, 'would probably be exercised, as in the largest living Saurians, in delivering deadly strokes on land, as well as in cleaving a rapid course through the watery element'.[8] The whale-lizards were slowly, if hesitatingly, emerging on to land. The upper arm bone was massive which suggested the name of the dinosaur: Owen called this 'cetiosaur' *Omosaurus*. *Omosaurus* means 'humerus saurian'. The bones in the hand were short and stumpy, while the 'whole fore-foot', observed Owen, 'was more massive and elephantine in its proportions' than in any other Mesozoic giant: the enormous feet must have been used for terrestrial locomotion. *Omosaurus* was not, as it turned out, a cetiosaur, nor even a close relative, but rather a smaller dinosaur related to the plated *Stegosaurus*. What is important is that Owen *thought* that it was a cetiosaur and acted on that belief. The presence of five fused hip vertebrae, reasoned Owen, was positive evidence that cetiosaurs were dinosaurs.

But Owen had been anticipated. His extreme caution and strict adherence to his own ageing definition of the dinosaur (which others had begun to view more flexibly) allowed his younger London contemporaries to move ahead. In particular, Huxley, the perennial thorn in Owen's side, was not slow to sieze the opportunity. The two palaeontologists had broken diplomatic relations in 1856, after Owen had assumed Huxley's title of Professor of Palaeontology while delivering a lecture course at Huxley's Royal School of Mines. The following year Huxley retaliated by defeating Owen in the academic arena. The issue was of course man; the last bastion to the pious. Owen, in his 1857 Linnean Society lecture, had set man apart from the rest of Creation on anatomical grounds. If God was to remain a viable proposition, if man was to keep his soul, then man must be *seen* to be unique, so Owen attempted to demonstrate that God had bestowed certain unique features upon man's skull and brain as a sign of his special favour. With the earth no longer holding its privileged position in the cosmos, with *Genesis* no longer the yardstick by which chronology was measured, then at least man would remain inviolate. Huxley took great delight in knocking the bottom out of Owen's arguments by pointing to these so-called unique features in the higher apes. The last bastion had fallen. Huxley now turned his attention towards the dinosaurs, and again challenged the Old Man of palaeontology. In 1867 Huxley visited the Oxford Museum at Phillips' invitation and marvelled at Buckland's cetiosaur collection. There could be no question but that these monstrous creatures were dinosaurs, asserted Huxley, and in 1869 he published a new classification of the dinosaurs to include Phillips' *Cetiosaurus*, allying it specifically with the iguanodonts.[9] Such a union was hotly disputed by Harry Govier Seeley, although he too had no doubts about the dinosaurian nature of the giant cetiosaurs. Seeley, in fact, realised that cetiosaurs were a major and distinct group of dinosaurs. Perhaps, he ventured, 'distinct original groups are confounded under the name Dinosauria'. Dinosaurs were an important group of animals in Mesozoic times, but they were made up of major sub-groups whose differentiation Seeley now proposed. One of the groups he immediately separated off was the Ceteosauria and in 1874 he suggested that they be raised in rank to an order.[10] The whale-lizards had begun to acquire some importance. Seeley's fruitful line of reasoning and his search for some criterion to distinguish one type of dinosaur from another gave way to a more natural classification of the dinosaur in the following decade.

For nearly forty years the cetiosaur had been seen as aquatic, jousting for dominance in the Mesozoic oceans with the ichthyosaurs and plesiosaurs. But massive pachydermal limbs and stumpy toes had shown how untenable that view had become. Still, the complete jump from water to land was a hard step to take in one go. Although the enormous cetiosaur possessed the 'faculty of terrestrial progression in a superior degree', Owen struck a compromise between 'whale-lizard' and dinosaur by his reaffirmation that 'the habitual element of the Cetiosaur may have been, and I believe to have been, the waters of a sea or estuary'.[11] Owen was unable to make the break from traditional ideas. Even though his monster was endowed with massive limbs to support its weight, Owen never really allowed the creature to leave the water that had been its home for so long. This despite his earlier plea that the dinosaurs, unlike the marine plesiosaurs and their kind, were by definition land creatures. Hidebound by tradition, Owen left his monstrous 'whale-lizard' swimming in the shallows. But he had, by the 1870s, made significant advances in man's knowledge of the dinosaur by including within its domain giant four-footed sea and lake dwellers.

The English cetiosaurs were an ill-defined assemblage in Owen's day and knowledge of them has not greatly increased since. But however painful had been their birth, the really giant dinosaurs had made their Victorian debut. These protracted birth-pangs were hardly experienced by the New World; here the monsters sprang fully formed from the earth. It was as though a veil was dramatically lifted from the creature. In 1877 O. C. Marsh at Yale College received from the West the first fossil bones of the cetiosaur's American cousin; no longer would palaeontologists be forced to cope with crumbling and fragmentary remains, from now on they were to unearth whole specimens perfectly preserved in a veritable giants' graveyard.

That such giants should ever have seen the light of day owes more to the vagaries of human motivation than dispassionate scientific study. The bones that Marsh received had been stumbled upon by Arthur Lakes, an Oxford graduate who had taken a teaching post in a Denver mission school. Having acquired a deep interest in natural history at Oxford, whose Museum was a repository bursting with saurian remains, Lakes had been investigating the virgin territory around his school. One day in March, 1877, while out looking for fossil leaves in the hard Dakota sandstone near the town of Morrison, Lakes' eye fell upon a giant vertebra embedded in block of stone. Since he was familiar with Professor Phillips' 'whale-lizards', he wrote to Marsh at Yale and included a sketch of the vertebra and an arm bone. Marsh's reputation had already been established by his Kansas pterodactyl and fossil bird finds, which made him a natural target for news of such discoveries. Lakes thoroughly explored the site and soon unearthed an enormous thigh bone. So again he wrote excitedly to Marsh, estimating that the monster must have been not less than sixty or seventy feet long. No reply came from New Haven. Lakes sent off a third letter, followed by a fourth. He told of other colossal bones belonging, he said, to six individuals if not six different species. Quite unbelievably Marsh was not interested. Lakes packed the bones and waited several weeks for Marsh's signal to send them, but still no word arrived. He persevered and in May, 1877 he took it upon himself to ship ten crates full of dinosaur bones, whose total weight was little less than a ton, to New Haven. Marsh's procrastination finally defeated Lakes. In desperation he took a step that was to make Marsh sit up and take notice – by writing to the one person that Marsh most feared laying his hands on prize specimens. Lakes wrote to Cope. Marsh had waged a bitter, if somewhat defensive, campaign against Cope since the days of the elasmosaur quarrel and was constantly repulsing his opponent's insidious attacks in the press. Now Lakes had not only written to Cope but he had actually shipped some of his giant bones to Philadelphia. Marsh got wind of this correspondence and was immediately spurred into action, dispatching a cable to Lakes to prevent him (belatedly) sending any of his Morrison bones to Cope, and a telegram to one of his own field collectors in Kansas, Benjamin Mudge, ordering him to Morrison at once to strike a deal with Lakes and secure what Marsh now regarded as his property. Arriving in Morrison, Mudge was quickly made aware of the rich collecting to be had as Lakes 'trotted out the menagerie' to him. An agreement was reached and Mudge cabled Marsh: 'Satisfactory arrangement made for two months. Jones cannot interfere.'[12] The 'Jones' was, of course, Cope. Cope meanwhile had been hastily preparing a paper on these saurian remains and had delivered a preliminary account to the American Philosophical Society in Philadelphia. It is not hard to imagine his anger as they were snatched back by Lakes, who informed Cope that Marsh had already purchased them. To add insult to injury, Lakes requested Cope to send his fossil bones on to Marsh.

Even a cursory glance at Marsh and Cope at war quickly explodes the myth of the sober, objective scientist. The jealous rivalry that existed between these two men in the days of the American West was not without its compensations, for it was undoubtedly the most productive quarrel in the annals of palaeontology. Both men were prolific describers and their motivation was often that the other was working the same area. How long would it have taken Marsh to wake up to Lakes' discoveries if Cope had not suddenly acted as a catalyst?

Marsh lost no time publishing an account of Lakes' fossil bones in order to secure his priority over Cope while, in Morrison, Mudge began working the quarry in earnest. The first bone he extracted, an upper arm bone over two feet wide, foreshadowed things to come. Within a few weeks Lakes and Mudge had shipped another ton of bones by rail to Marsh, among which were the first known remains of the plated dinosaur *Stegosaurus*. At first Marsh was not too sure of the stegosaur's affinities and thought that it may have been related to the turtles, seeing the animal's huge dermal plates (some of which were over three feet long) as a sort of tank-like armour plating.[13] Later he up-ended these plates to form a tall bony crest running down the midline of the stegosaur's back.

The size of some of the bones in Marsh's early shipments is quite staggering even to us today. Most of the early discoveries were of *Titanosaurus*. One thigh bone in Marsh's possession was over eight feet long; if such a bone had been owned by a crocodile, speculated Marsh, the creature would have been an unbelievable one hundred and fifteen feet long! Like its English cousins *Titanosaurus* had hollowed vertebrae, but only three or four vertebrae comprised its fused hip region (less than in Owen's original definition of the dinosaur). When this region became better known in dinosaurs the number was found to be variable among different types.

By a peculiar twist of fate, the sort that seems to have doggedly followed Marsh and Cope for most of their working lives, another schoolmaster was roaming the Colorado hills in March, 1877. O. W. Lucas, an amateur botanist, was scouring the region near Cañon City to the south of Morrison in search of plants. He, too, came upon huge saurian remains but *his* initial reaction was to write to Cope. Cope seized the opportunity to draw level with Marsh and promptly engaged

Lucas. Excavations began in haste. The first bones shipped back to Cope in Philadelphia consisted of fragments of a *Laelaps* jaw. But the second batch were vertebrae which, announced Cope, 'apparently represent a much more gigantic animal, and I believe the largest or most bulky animal capable of progression on land of which we have any knowledge'.[14] (It is hardly surprising that these words were to be uttered with monotonous regularity during the following months as larger and larger creatures emerged from the quarries.) Cope rushed out a paper on August 23, 1877 describing Lucas' giant reptile, which he named *Camarasaurus* or 'chambered reptile' after the cavernous nature of its vertebrae, although he had been narrowly beaten by Marsh, whose paper on *Titanosaurus* had appeared on June 20. But Cope had a remedy for being a little late in publication. He insisted that *Titanosaurus*, the name chosen by Marsh, had no claim to be adopted according to the rules of nomenclature because there was no accompanying diagnosis or descriptions of the bones. From Marsh's words it would have been impossible to distinguish the creature from an English cetiosaur. Anyway, added Cope for good measure, *Titanosaurus* was a name already taken by a fossil creature. Cope, coming in a poor second to Marsh, was going to contest the decision on legal technicalities. It was, Cope implied, *he* who had first described a giant herbivore. In deference to the rules of nomenclature Marsh *did* change the name of his monster to *Atlantosaurus*, although this is little justification for robbing him of priority.

Until this time Cope had been dealing solely with two-footed kangaroo-like dinosaurs, so it came as some surprise to him that the front feet of *Camarasaurus* were long and that the creature moved on all fours. Cope's initial reaction on learning this is surprising. The long fore limbs, he remarked, 'taken in conjunction with the remarkably long neck possessed by that genus, suggests a resemblance in form and habits between those huge reptiles and the giraffe. While some of the later *Dinosauria* elevated themselves on their hind limbs to reach the tree-tops on which they fed [he is referring to the Cretaceous *Hadrosaurus*] the general form of the body in some of these earlier types enabled them to reach their food without the anterior limbs leaving the earth'. Lucas' next shipment included the remains of a still taller beast, *Amphicoelias*, which Cope saw as a fit rival to the *Camarasaurus* in the Mesozoic world. Cope's unexpected reversal of the lifestyle of the dinosaur came barely two years after Richard Owen in London reaffirmed his belief that these gigantic quadrupeds were aquatic. It appeared as though the cetiosaur's American cousin had suddenly been emancipated from the sea. But the leviathan was struggling ashore prematurely.

Marsh, meanwhile, got wind of his rival's dig only a hundred miles to the south of his own and cabled Mudge to leave Morrison and investigate. Cope, however, had already monopolised the site and Lucas was in his pay. The bones that they were digging out, noted Mudge disappointedly, were in better condition and anything up to 30% larger than their Morrison counterparts, and it was costing Cope very little to excavate them. Marsh ordered his collectors into the region to begin prospecting nearby, whilst Samuel Williston was brought from Kansas to direct operations with Mudge. When they sensed that Lucas was disillusioned with Cope, feeling himself exploited and underpaid, overtures were immediately made to him, but Lucas honoured his agreement with Cope and shipped the remainder of the bones in his quarry to Philadelphia. Mudge and Williston's alternative quarry failed; they were removing only friable and broken bones from the hard sandstone and they jealously eyed Cope's men picking out complete bones from the softer shale in their quarry. After the failure to buy off Lucas, Marsh ordered his men to pull out and return to Morrison to collect there as best they could, but a serious rockfall almost killed the entire party and work was suspended for the rest of the season.

Again it transpires that Marsh did, in fact, hear of the Cañon City site before Cope. Letters from yet another collector were sent as early as February, 1877 reporting the discovery of giant reptilian bones, some with articulating heads twenty inches in width. In this instance, too, Marsh had failed to follow up the tip and it was only when Cope moved in that Marsh decided that the site was of scientific value.

By mid 1877 Marsh had become fully alerted to the fantastic wealth of fossil reptiles waiting to be excavated in the Jurassic rocks of the Rocky Mountain states. In July he received a letter penned under the pseudonym of 'Harlow and Edwards' claiming the discovery of large numbers of fossils, supposedly those of the giant ground sloth *Megatherium*, in rocks presumed by the discoverers to be of Tertiary age. 'We are desirous of disposing of what fossils we have,' wrote Harlow and Edwards, 'and also, the secret of the others.' One shoulder blade, they informed Marsh, measured a little under five feet. 'We have said nothing to anyone as yet,' they confided. 'We would be pleased to hear from you, as you are well known as an enthusiastic geologist, and a man of means, both of which we are desirous of finding – more especially the latter.' As a sign of good faith they shipped some bones to New Haven for Marsh's inspection.

Marsh unpacked the giant sloth to be rewarded by vertebrae, teeth, claws and limb bones of the huge Jurassic *Brontosaurus*. He immediately telegraphed Harlow and Edwards and forwarded a cheque which, of course, could not be cashed because the discoverers had concealed their real identity to keep the find and its location a secret. They wrote back to Marsh by way of an explanation: 'there are plenty of men looking for such things and if they could trace us they could find discoveries which we have already made'. Marsh urgently cabled Williston and ordered him to this new quarry, located at Como Bluff in Wyoming. Williston let it be known that he was going to Oregon, just in case anyone should try to follow him. (Such events read more like a fictional spy drama than fossil exploration; secrecy and even deception were symptomatic of the paranoia that had overtaken the rival camps in their quest for giant saurians.) The bones were scattered for *six or seven miles*, wrote Williston of this new locality, 'Cañon City and Morrison are simply nowhere in comparison'. He wasted no time signing up the two discoverers, Como railroad men William E. Carlin and Bill Reed, and within three days formalities had been settled, before even Cope had got wind of the deal. The agreement even included Carlin and

Reed taking 'reasonable precautions to keep all other collectors not authorised by Prof. Marsh out of the region'. A justifiable precaution, of course. Within days a mysterious character made repeated appearances around the site, inquiring after bones and skulls. One of Cope's men, reported Williston. 'He first purported to be selling groceries!!' Williston even obtained a sample of his handwriting for comparison with Cope's, just in case it was March's arch rival in person.

Thus began exploration at Como Bluff, where one site gave way to perhaps a hundred dinosaur-yielding quarries in very few years. The intrigues continued. The two palaeontologists developed rapid dinosaur extraction into a fine art, hoping to clear out a quarry before the other was aware of its existence. Ugly incidents were rife, men deserted to the opposite camp and strong-arm tactics were employed to drive invaders out – or move into the rival quarry. Huge dinosaurs like *Diplodocus*, *Camarasaurus* and *Barosaurus* became a common feature of Marsh and Cope's scientific papers. Quarries were opened, stripped of all their bones and closed all within the space of a few months, or even weeks. When the party left, they smashed any remaining bones to stop rivals laying their hands on them. Such exploitation brought to light many fossils that may have gone unnoticed by any other approach. In one quarry, much excitement was generated by the discovery of the first American Jurassic mammal, a primitive 'marsupial' creature thought to resemble an opossum, which a delighted Marsh named *Dryolestes*. (In point of fact, it lived far too early in time to have been a marsupial; the marsupials and placentals did not appear until mid Cretaceous, having arisen from primitive mammals related to the dryolestids.) A spectacular lure such as this brought Cope himself to the quarries in search of similar trophies.

As early as 1878 the influx of giant herbivores' bones to New Haven was so great that Marsh singled them out for special study, devoting several papers to them. Like Seeley before him, he acknowledged the giant four-footed dinosaurs as somewhat distinct from the others and officially set them aside as the Sauropoda or 'lizard feet' on account of their five-toed feet (in contrast to the bipeds which were usually three-toed).[15] Although Seeley had already distinguished these dinosaurs as the Ceteosauria, it is Marsh's term that has passed into common usage.

The digging continued over the years with Marsh and Cope striving to keep pace with one another. Paper after paper was delivered and published describing the creatures that emerged from the rocks. Whole brontosaurs were brought to light, plated stegosaurs, giant carnivores such as *Allosaurus*, and even cat-sized dinosaurs like *Nannosaurus*. Large numbers of bones of the sauropod *Morosaurus* (Cope's *Camarasaurus*: because the two palaeontologists were describing the same dinosaurs simultaneously most of the creatures ended up with two names) were unearthed. When Marsh first saw the ridiculously small skull on this monstrous creature he was astonished; its brain must have been relatively smaller than in any other backboned animal. The large bipedal ornithopod *Camptosaurus* was found in these quarries, as well as the delicate *Coelurus* with hollowed bones and bird-like features (a peculiar creature that Marsh could not at first place although by the early 1880s he recognised it as a type of carnivorous dinosaur). The bonanza lasted twelve years. By 1889 the great bone rush was over. The quarries were worked out and very few have yielded comparable giants subsequently.

One outcome of the discovery of whole sauropods at Como was that Marsh was able to accurately reconstruct these creatures for the first time. The *Brontosaurus* skeleton found by Reed was almost perfect; all the bones had come from one individual and this must have been fifty feet long in life. Marsh restored this skeleton in 1883 and accompanied his restoration by a description of the creature.

> The head was remarkably small. The neck was long, and, considering its proportions, flexible, and was the lightest portion of the vertebral column. The body was quite short, and the abdominal cavity of moderate size. The legs and feet were massive, and the bones all solid. The feet were plantigrade, and each foot-print must have been about a square yard in extent. The tail was large and nearly all the bones solid.
>
> The diminutive head will first attract attention, as it is smaller in proportion to the body than in any vertebrate hitherto known. The entire skull is less in diameter or actual weight than the fourth or fifth cervical vertebra.
>
> A careful estimate of the size of *Brontosaurus*, as here restored, showed that when living the animal must have weighed more than twenty tons. The very small head and brain, and slender neural cord, indicate a stupid, slow moving reptile. The beast was wholly without offensive or defensive weapons, or dermal armature.
>
> In habits, Brontosaurus was more or less amphibious, and its food was probably aquatic plants or other succulent vegetation. The remains are usually found in localities where the animals had evidently become mired.[16]

It was not blind acceptance of traditional views that caused Marsh's preference for an aquatic sauropod. There was one feature that these dinosaurs shared with truly aquatic creatures like ichthyosaurs and whales: the nostril was placed far back on the skull, often lying near to or even between the eyes rather than on the tip of the snout. An aquatic creature is in danger of shipping water when it breathes, so it is desirable to have the nostril as high as possible on the skull. The nostrils were even placed on domes in *Brachiosaurus*, so they could be raised above the water whilst the rest of the animal remained immersed. Unlike Marsh, Cope's initial reaction to *Camarasaurus* was that he was dealing with a reptilian giraffe, endowed with a wonderfully elongated neck enabling it to browse at tree-top level. In accepting this Cope had willingly brought the animal ashore for the first time. Yet he soon abandoned this idea and threw his weight behind the consensus opinion that sauropods were lake dwellers, picturing them totally immersed and standing on the firm lake bottom, or wandering under water in search of food. Occasionally they would raise their elongated necks to bring the head above water to gulp down air before continuing to pluck the succulent vegetation from the lake floor.[17] At even greater depths the animal, buoyed up by the water, would rear up on its hind legs to reach the surface some thirty or more feet above the lake floor. The huge sauropod lounging on the lake bottom out of

39. Cope's snorkelling *Amphicoelias*. The notion of submerged sauropods, feeding on lake floor vegetation but rising periodically for air, has remained dominant until the present day.

harm's way, raising its periscope-like neck vertically to break surface, has remained a popular notion since Cope's day. Cope's biographer, Henry Fairfield Osborn, agreed wholeheartedly with his predecessor. He imagined *Camarasaurus* as a great quadruped wading across the firm sandy river bottoms of a Jurassic Wyoming, or perhaps swimming rapidly when danger threatened by propelling itself along with a long powerful tail. An extensive cartilaginous casing to the limb joints and the supposed cartilaginous nature of some ankle and wrist bones were thought by many dinosaur experts to mean that the sauropods could never have moved on land without grinding their limb joints and damaging themselves. Osborn did not subscribe to this extreme view but, nonetheless, he too pictured *Amphicoelias* living mostly under water where the flexible emu-like neck was capable of twisting and turning, rising vertically every few minutes to take in air. This 'hypothesis of function', said Osborn, 'applied to all the Cetiosauria, namely, to the *Morosaurus* and *Diplodocus* types as well'.[18]

The brontosaurs known to the late nineteenth-century palaeontologists had snake-like necks about eighteen feet long crowned by ridiculously small skulls with dorsal nostrils. It was assumed that these acted in the manner of supersnorkels, permitting the reptiles to continue their submarine existence unhampered. But the earliest sauropods discovered, though enormous by any standards, were by no means the largest snorkelling dinosaurs to have lived. Marsh's *Brontosaurus* was, in fact, only fifty feet long. Osborn was working with a 'large individual' of *Camarasaurus* unearthed by an American Museum party in 1897, a creature that probably reached sixty feet or so.

After Marsh had described the first few isolated bones of *Diplodocus*, high hopes of discovering a really good specimen were shattered on every field trip. Osborn's case was typical. He found the tip of a perfectly preserved *Diplodocus* tail and proceeded to follow it, vertebra by vertebra, by cutting back the cliff in which it was embedded. But instead of the bones getting larger and more numerous, they petered out after only a few feet. Then the Carnegie Museum in Pittsburg entered into the race for these prestigious giants, and it was the museum's collecting parties of 1899 and 1900 that finally brought back the first imperfect but reconstructible skeletons from Albany County, Wyoming.

These forays to the West were sponsored by one man, both tyrant and hero in America's Gilded Age, Andrew Carnegie. Carnegie had risen through the ranks from telegraph messenger boy to multimillion dollar steel tycoon. In an age when steel was king and Capitalism knew none of its latter-day refinements, Carnegie's ruthless exploitation of man and earth seemed totally in keeping with the prevailing ethic, an ethic that Carnegie found to his taste. Carnegie justified his race against competitors, his insatiable demand for higher production, and his sacrificial offering of workers and colleagues alike on the high altar of efficiency as an execution of the Social Darwinist blueprint. Herbert Spencer's brand of Social Darwinism found a ready advocate in post Civil War American Business and Andrew Carnegie in particular: entrepreneurial competition, in weeding out the weakest elements of industrial society, would ensure stronger survivors. The new Capitalist ethic, which many saw elevated into a law of nature, was Carnegie's credo. (Herbert Spencer remained detached from the harsh world depicted in his writings, preferring the insularity of his English home; when Carnegie eventually enticed him to Pittsburg, eager to show off his industrial Utopia founded on true Spencerian principles, the deflated English philosopher turned to Carnegie sorry-eyed and lamented: 'Six months' residence here would justify suicide.'!)[19] As an apologist for accumulated wealth, Carnegie was articulate and forceful: he did not regard himself as its *owner*, merely its *custodian*. And much to Spencer's disgust, Carnegie derived even greater satisfaction from redistributing his fortune. Indeed, in his *Gospel of Wealth*, Carnegie allotted pride of place to its wise disposal (he was motivated by the belief that his endowments would act as a stimulus to others to strive as he had done). Thus in 1895 he founded the Carnegie Institute in his home town of Pittsburg and soon after built and equipped a museum. Carnegie poured $25 million into the Institute in its first ten years, aid which secured for Pittsburg some of the best sauropod skeletons ever found. But the most celebrated by far was *Diplodocus*.

J. B. Hatcher, the Museum Director, attempted a paper restoration of *Diplodocus* in 1901 based on the latest Wyoming finds, but with the gaps filled by bones from many other collections. The resulting creature was *thought* to be a little under sixty-five feet long but that, as it turned out, was a gross underestimate. The strange proportions of *Diplodocus* were immediately apparent from Hatcher's restoration. It had a remarkably long neck containing over fifteen elongate vertebrae, a tail with over thirty-seven vertebrae and a relative short trunk in between. Since the Carnegie specimen differed from previously known species, it was christened *Diplodocus carnegiei* in honour of the Museum's patron. Carnegie was delighted with this and refused to settle for a mere paper restoration, charging the preparators with the task of a full-scale skeletal restoration at his expense. Meanwhile new specimens of Carnegie's *Diplodocus* in far better condition had been discovered in Wyoming and these were incorporated into the new restoration by Hatcher's successor W. J. Holland. A complete skeleton, made out of plaster of paris, was constructed and housed in the Carnegie Museum. Many of the vertebrae, it transpired, had been missing from Hatcher's earlier restoration. The tail did not possess thirty-seven vertebrae; Holland now had *seventy-three*, and articulating facets on the last of these suggested that the tail's tip had still not been reached.[20] The restoration of *Diplodocus carnegiei* in the Carnegie Museum was over eighty-seven feet long. At almost thirty feet longer than Marsh and Cope's earlier finds, it was by far the largest land animal known. And relative to its size it had one of the smallest brains; an immense machine directed by a minute control centre, much as the Pittsburg steel empire was dominated by a single man (an analogy favoured by Carnegie's biographers).

One former Trustee of the British Museum who could be counted on to wield no little influence over the Pittsburg industrialist was King Edward VII. The English monarch had been staying with Carnegie and had marvelled at his restored *Diplodocus*, casually expressing the hope that one day the British Museum might be lucky enough to acquire such a monster. Carnegie immediately set his preparators to work, taking another set of casts from his *Diplodocus* moulds. Within a short time a complete duplicate of the Pittsburg dinosaur had been made and it duly arrived at the British Museum in January,

40. Holland's *Diplodocus carnegiei*, over 87 feet long, and standing erect like a mammal.

1905. In April Carnegie's chief preparator also arrived in London and took control of the reassembly from the three hundred or so bones. The creature was so large that no room could be found for it in the Hall of Palaeontology, as it then was, so it was mounted in the Gallery of Reptiles (later to become the Dinosaur Gallery). The Trustees arranged a formal handover on May 12 and the Gallery was packed with eminent scientists and palaeontologists (with a liberal sprinkling of lords) to see the occasion through. Professor Ray Lankester, the Director of the Museum, introduced the proceedings, whilst Lord Avebury accepted the *Diplodocus* from Andrew Carnegie on behalf of the Trustees.[21] The event obviously impressed Carnegie because he sent Holland ahead as his emissary to the monarchs of other European countries, and his *Diplodocus* was shortly presented with great pomp to Kaiser Wilhelm and the President of the French Republic. Casts also found their way into the museums of Vienna, Bologna, La Plata in Argentina and Mexico City. In each city Carnegie's preparators accompanied the crates of plaster dinosaur bones in order to mount the creature, and Holland remained just long enough to supervise the installation and act as Carnegie's representative at the presentation ceremony. This same *Diplodocus* is easily the most impressive exhibit in many of the world's finest museums. Each of the skeletons is reputed to have cost Carnegie $30,000 to make up, ship to its destination and then be reassembled and mounted. With such splendid promotion it is hardly surprising that *Diplodocus* is one of the best known inhabitants of the ancient world.

All this time, nobody thought to question the upright stance of the mounted sauropods. *Diplodocus* and its kin had always been restored in an elephant-like pose, standing with legs tucked under the body. Like an elephant, the legs functioned as props to support the body off the ground. Marsh's first restoration of *Brontosaurus* in 1883 set the trend, depicting the creature walking with its body high above the ground on straight pillar-like legs. 'Marsh's example has been followed slavishly ever since,' commented a dissident Oliver P. Hay in Washington in 1908. In fact, the *Brontosaurus*, although one of the largest and

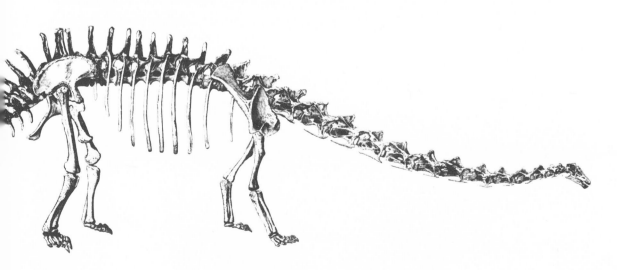

41. The formal presentation of Carnegie's *Diplodocus* to the British Museum in 1905.

heaviest beasts of all time, was mounted in the American Museum with a slight curve in its legs, making it appear bow-legged. But the *Diplodocus* that Carnegie had built was perfectly straight-legged.

Why was an elephant pose first employed? The two-footed dinosaurs were the first known in America and they completely dominated dinosaur studies for many decades. When Leidy and Cope came upon fossil creatures with huge hind legs, they could immediately see a parallel in the present-day fauna. This was not to be found among reptiles, however, but among the mammals. So it is understandable that they should use the only model at their disposal and restore *Hadrosaurus* and *Laelaps* kangaroo-style. So ingrained had this picture of the dinosaur become that when Cope found the first sauropod giants like *Amphicoelias* he automatically assumed that they also walked erect on their hind legs. With the discovery of equally massive fore limbs he was forced to drop the creature on to all fours. Since bipedal dinosaurs like *Hadrosaurus* were capable of walking on their front legs – when the limbs would have been tucked under the body as they are in kangaroos moving on four legs – Marsh and Cope used this as their model. They merely lengthened the fore legs and adopted the resulting elephant-like pose. With forward facing legs brought under the body, Marsh's sauropods had the stance that Richard Owen had bestowed on his London beasts many years earlier.

When Carnegie distributed his *Diplodocus* specimens (and miniature scale models) around the world scientists for the first time came face to face with a reptile that looked like a pachydermal mammal. With the arrival of *Diplodocus* in London a well-known English illustrator, Mr F. W. Frohawk, clearly saw the paradox in Carnegie's restoration, and voiced his doubts in *The Field* for 1905:

> The visitor to the Reptile Gallery of the Natural History Museum cannot fail to be struck by the extraordinary pose of the gigantic skeleton and miniature model of the *Diplodocus carnegiei*. It would be interesting to know the reason for mounting the specimen so high on its legs, like some huge Pachyderm. As it is a gigantic lizard, why should it not be represented in the attitude usually assumed by such animals? The skeleton has the appearance of belonging to some great mammal. The small model made by Dr Holland shows us a creature with body and limbs resembling an elephant, with enormous length of neck, and tail like an immense python. It appears reasonable to suppose that an animal with a neck of of such prodigious length would hardly carry it in the manner represented, but that the greater part of the neck, body and whole of the tail would more likely rest on the ground, and limbs more lateral like other reptiles. Doubtless there is some good reason for mounting and modelling it in such an attitude; if so, information on the subject would be welcome.[22]

The following year there was an attempt to reconstruct just such a crawling dinosaur, in this case *Brontosaurus*. Two assistants, Otto and Charles Falkenbach, in the palaeontological laboratory of the American Museum built a small model of *Brontosaurus* on crocodilian lines. The limbs of the dinosaur were directed outward so that the creature appeared to creep along with its belly just off the ground. This model was exhibited at the meeting of the American Society of Vertebrate Palaeontology held in the Museum that year but by common consent it was thought to represent the impossible.

Carnegie's *Diplodocus* must take responsibility for precipitating the revolt proper, a revolt that was to become as truly international as the *Diplodocus* itself. While Holland was making the presentation to the French President, Hay – ironically, himself the recipient of a grant from another of Carnegie's beneficiaries, the Carnegie Institution of Washington – challenged the choice of pose for this creature (independently, it seems, of the American Museum assistants). It is highly improbable, he stated, that *Diplodocus* ever stood erect like an elephant. Hay's contemporaries had used the animal's enormous weight as an argument *for* upright legs. With pillar-like columns to prop up the body the weight is transmitted directly through the legs to the ground. The weight of an animal increases as a cube of its length whilst strength varies only as the cross section or square. With increasing size the dinosaur's weight eventually overcomes its strength to remain standing – especially if its legs are bent. Developing straight supporting columns is the most efficient means of sustaining this enormous weight. Hay, however, maintained that excess weight was an argument *against* a mammal-like carriage. The dinosaur was not a mammal but a reptile, and reptiles do not walk like elephants, they crawl like alligators. 'The conception of a creeping dinosaur was hardly to be entertained,' lamented Hay. Nevertheless, with all that weight to support, *Diplodocus* must have had sprawling legs so that it could spend most of its time flopped on to its belly like an alligator. The feet, he supposed, were directed outwards rather than forwards, whilst the knee and elbow were held at right angles as in a lizard. If the *Diplodocus* inhabited marshy country, moreover, it would be better equipped if it slithered like a crocodile. Being so heavy, if it walked erect over swampy terrain it would soon have become inextricably mired and have perished miserably. He continued:

> It is difficult to understand why an animal whose immediate ancestors must have walked about in a crocodile manner, an animal that was stupid and probably slow of movement, an animal which could by means of its long neck reach up from the bottom many feet to the surface and from the surface many feet to the bottom – why such a reptile should need to develop the ability to walk along river bottoms like a mammal.[23]

Diplodocus was a lazy creature (not by choice, but through necessity) and could creep about on land only with great effort. Its real home was a swamp. Of course, a crocodilian *Diplodocus* lacked some of the novelty attached to an elephantine one, but this could be remedied:

> It seems to the writer that our museums which are engaged in making mounts and restorations of the great Sauropoda have missed an opportunity to construct some striking presentations of these reptiles that would be truer to nature. The body placed in a crocodile-like attitude would be little, if any, less imposing than when erect; while the long neck, as flexible as that of an ostrich, might be placed in a variety of graceful positions.

42. Hay's 'creeping' *Diplodocus*, restored to crocodilian specifications.

Hay brought *his Diplodocus* to life in a line drawing in 1910.[24] To a generation reared on elephantine sauropods, the scene he conjured up appears strange indeed, although probably no stranger than Carnegie's elephantine monster appeared to its first English visitors.

With the spread of the plaster *Diplodocus* across the world, the controversy raged in its wake. In Vienna Othenio Abel was quite happy with the mammal-like pose, whereas in Germany Gustav Tornier sided with Hay, preferring a crocodile to a mammal. Since *Diplodocus* was a reptile, he reasoned, it should never have been restored to its former glory looking like a mammal, and as proof that a crocodile pose was not impossible he drew a sketch restoration. Other German palaeontologists followed Tornier's lead and mocked their American counterparts for misconstructing the creature from head to toe. This was too much to take and American palaeontologists, seeing themselves held up to ridicule, poured out counter-polemics directed against the German 'closest-naturalists', as Holland styled them. With the help of a pencil, said Holland, Tornier had dislocated joints and bent bones, squeezing the dinosaur into a

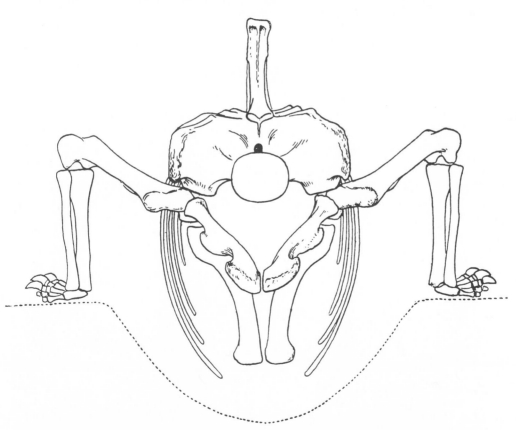

43. A section through the skeleton of the 30 ton *Diplodocus* restored in a lizard pose. The attempt to make *Diplodocus* a sprawling dinosaur was a dismal failure. As Holland sarcastically noted, the deep dinosaurian ribcage would have necessitated a rut in the ground if the animal was to crawl.

grotesque shape. His monstrosity would have been totally incapable of any motion whatsoever. Two decades of labour by leading American students of the dinosaur had been successfully wiped out in a stroke of Tornier's pencil. 'As a contribution to the literature of caricature the success achieved is remarkable,' was Holland's derisory assessment of Tornier's 'skeletal monstrosity'.[25] The only feature to recommend it, said one European palaeontologist, was that it accounts for the speedy disappearance of the sauropods. With every joint dislocated, their lives must have been spent in indescribable agony!

Carnegie's goodwill gesture had degenerated into an international slanging match, with not a little hurt pride on all sides. Holland's 'cruelly convincing polemic', as Matthew aptly termed it, effectively silenced Tornier and his followers. Holland's broad knowledge of both dinosaurs and modern reptiles was second to none. A detailed study of the joints of the limbs, girdles and feet led him to pronounce quite categorically that *Diplodocus* was 'rectigrade', walking erect on forward-facing pillar-like legs. The massive size of dinosaurs like these had necessitated such colossal supporting limbs.

Holland was completely vindicated in the late 1930s when Roland T. Bird chanced upon the first sauropod tracks. In Glen Rose, Texas, in 1938 he found a complete trail left in the Cretaceous mud after the passing of a sixty-foot brontosaur. The monster's stride was a good twelve feet, yet the width of the rail was only six feet.[26] If the brontosaur had been crawling like a crocodile with legs sprawling at the sides of its body, the width would have been much greater. Here was unequivocal evidence that brontosaurs and their kin walked staunchly upright.

The controversy had faded into history. From our alternative perspective, aided by a new set of presuppositions underpinning our view of the dinosaur, we can appreciate the paradox met with by our predecessors. Reptiles cannot maintain a high energy output for long and are consequently forced to rest on their bellies for most of their lives. This even applies to a small lizard of negligible weight. Hay's problem was to explain how a creature whose weight was vastly superior to a lizard's could manage to remain erect. The largest dinosaurs, such as *Brontosaurus*, *Diplodocus* and *Brachiosaurus*, weighed anything from thirty to fifty tons when full grown (estimates of live weight vary considerably). Hay would have been right in believing that the dinosaur could never have remained standing if it was a reptile like a lizard. But it was not. The sauropods must have been warm-blooded and have had a high energy producing metabolism, which would have provided them with the sustained energy output they needed to remain on their feet carrying such weight. Constant effort is required to keep the body off the ground and the energy available to a dinosaur must have been far in excess of that available to a lizard or crocodile. The dinosaur adopted a mammal-like pose because it had a mammal-like physiology. It could not have been otherwise.

The test of any good theory lies in its ability to tie up previously unrelated facts. We find that endothermy not only explains the stance of the dinosaurs but it also helps us understand why they grew so large. Dinosaurs were uninsulated (except for scales and bony scutes in some) and lost heat through their leathery skin. The larger the dinosaur grew the smaller its surface area became relative to volume. A larger beast thus loses relatively less heat than a smaller one. High metabolic rate and naked skin barred them from filling the microniche occupied by lizards and small insect-eating mammals. Lizards have low metabolic rates and can tolerate lower temperatures whilst mammals have an insulating coat of fur. The dinosaur could never have maintained a constant high temperature if it decreased in size beyond a critical level, it would have perished from exposure. Is it any wonder that no microdinosaurs have ever been found? They were forced to be giants.

44. Tornier's 'skeletal monstrosity'. With every joint dislocated, the mis-shapen *Diplodocus* must have endured life in indescribable agony.

Most dinosaur experts have assumed that in order to be able to carry their many tons of muscle and bone, the dinosaurs' pillar-like limbs must have received some help. If the elephant is the largest size our present-day terrestrial mammals can manage, asked Matthew, how could the dinosaur have grown so much larger? He supposed that, like the whale, the sauropod relied on water for buoyancy. With the weight taken off its feet, it was then able to grow to gigantic proportions.[27] *Diplodocus*, with a twenty-one-foot neck, was an exceedingly long sauropod. This beast could still have breathed when standing in water over thirty feet deep, and this is where it was usually placed. (In this respect, the globe-trotting *Diplodocus* was out-classed by *Brachiosaurus*, unearthed only a few years later in Tanzania in East Africa. Although not as long as Carnegie's showpiece, *Brachiosaurus* probably weighed over fifty tons in life, and the disproportionately long front legs combined with the sinuous neck to raise the head to a towering forty feet.) Like *Brachiosaurus*, *Diplodocus* had a nasal opening high up on its head and between its eyes. It was also considered something of a helpless creature, a reason of sufficient weight to leave *Diplodocus* in the water, where it was presumed the carnivores never ventured.

Diplodocus appeared to have only badly socketed rake-like teeth, and these only in the front of its mouth. Such an inefficient biting apparatus was thought to function by pulling up soft and succulent water weeds that the feet had torn from the lake floor. That, anyway, was how Hatcher saw the creature obtaining its food. He also pictured the surroundings in which *Diplodocus* found itself. In Jurassic times the region that is now Colorado, Wyoming and Montana probably resembled the tropical Amazon basin with numerous lakes and large rivers surrounded by a dense tropical vegetation with broad, level valleys subject to periodic flooding.[28] Here in the midst of the luxuriant foliage and in a hot steamy climate the huge *Diplodocus* spent its days in perpetual search of succulent vegetation. Unfortunately, the extraordinary wear on the teeth of most *Diplodocus* skulls is difficult to attribute to soft, aquatic vegetation. The tall, peg-like teeth, often an inch and a half long, are worn absolutely flat at right angles to their length, and since the teeth in the upper and lower jaws could not grind against one another we have to search for another explanation. If these dinosaurs were herbivores, as everybody supposed, Holland suggested that the wear might be brought about by the animals scraping their teeth on the surface of stones while uprooting plants. He was more inclined to believe, though, that the starchy matter in the trunks of the cycad trees so abundant in the Mesozoic jungles formed their staple diet. The peculiarly flattened teeth, he ventured to suggest, were produced as the animals gnawed at the trunks of trees in an attempt to get at the pith.

How these dinosaurs obtained enough food to meet the requirements of their huge bodies is perplexing, although the fact that they were so successful testifies to the lack of any problem in the sauropods' minds. Gustav Tornier suggested that they ate animal food and were not herbivores at all. This would indeed supply them with nutritious food in a far more compact form and presumably they would have needed to consume less to meet their daily intake requirements.

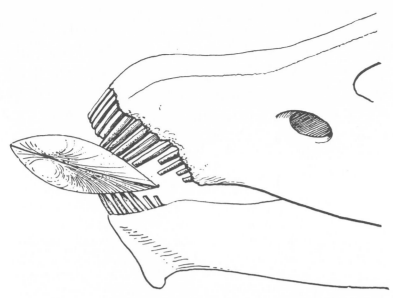

45. *Diplodocus* as a shell-fish eater. This would explain the flattened teeth; unfortunately, no traces of shells have been found in the stomach region.

Tornier's novel suggestion was that they were fish swallowers (they could just about manage fishes through their tiny skulls) although it was later proposed that they ate vast quantities of the bivalve molluscs which are known to have lived in profusion along the margins of Jurassic lakes. If *Diplodocus* wrenched shell fish off the lake bottom we are also able to explain the strange wear on the animal's teeth, a wear suggestive of the hard abrasive material in mussel shells. The *Diplodocus*, ventured Holland, may have seized the mussels in its front nipping teeth and uprooted them from the lake floor. It is an ingenious hypothesis. It would explain how a gigantic beast could have kept itself supplied with food through a tiny head, and why the teeth are worn in such a peculiar fashion. Unfortunately, despite extensive searches, Holland was never able to find clumps of swallowed mussel shells in that region of the fossil where the stomach was presumed to lie. He knew of several specimens of *Diplodocus* with vertebrae and ribs in place but no visceral contents were ever found. 'There is thus so far no confirmation of Tornier's theory,' sighed Holland, but adding, 'nor, I may say for that matter, of Professor Marsh's statement that these creatures were herbivorous. I confess finding myself compelled to announce myself as an agnostic, so far as the mode in which these creatures obtained their food.'[29]

Because the dinosaur resided in water, the food was always assumed to be aquatic. One or two dissenting voices were heard to cry out against this. Henry Fairfield Osborn at the American Museum noted that the vertebrae of the sauropods were ultra light, being hollowed-out and thus quite unlike those of whales. Brooklyn Museum had a whale skeleton seventy-five feet long and weighing almost 18,000 lbs. Osborn estimated that a sauropod skeleton of the same length weighed a little over half of this. 'Lightness of skeleton is a walking

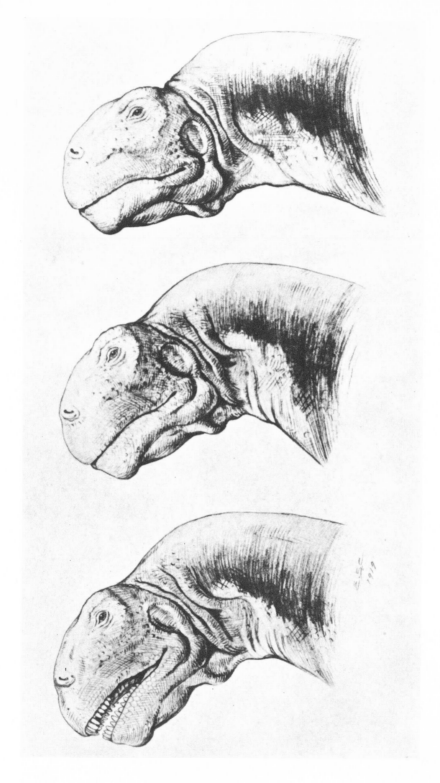

46. Conjectured facial expressions of Cope's *Camarasaurus*.

or running or flying adaptation,' he added, 'and not at all a swimming one; a swimming animal needs gravity in its skeleton, because sufficient buoyancy in the water is always afforded by the lungs and soft tissues of the body.' But Osborn was reluctant to take this line of reasoning to its logical conclusion and he settled for a compromise, declaring that the 'extraordinary lightness of these dinosaur vertebrae may therefore be put forward as proof of supreme fitness for the propulsion of an enormous frame during *occasional* incursions upon land.'[30] The dinosaurs were only allowed out of the water for brief periods, supposedly for egg laying or migration to a near-by pond. Osborn, like Cope in his earlier speculations, saw the animals stop to browse on tree-tops on their way to a new body of water and he even visualised them stretching up on their hind legs to reach the highest leaves. In such a position the herbivore could have defended itself against the smaller carnivores, rearing up when under attack to deliver a thunderous blow from its massive front limbs.[31] But this was not the sauropod's environmental preference: it was only in the water that the creature was really at home. Recognition that they must have come out into land to breed was an important step, yet even on the question of egg laying there was no unanimous agreement. Matthew argued that the sauropods may have been viviparous like ichthyosaurs, bringing their young into the world (and water) alive.

While out of water they were undoubtedly exposed to attack from the carnivores, as evidenced by the remains of at least one prehistoric feast. An American Museum crew turned up Jurassic brontosaur bones scored by tooth marks, and taking a lead from this, mounted their allosaurs hovering over the remnants of this scarred sauropod skeleton. Since the allosaurs have always been thought of as terrestrial, these feasts must have taken place out on dry land, where either the brontosaur died naturally or was brought down by the allosaurs.

Any lingering doubts concerning the brontosaur's amphibious habits were dramatically removed by Bird's discovery of a unique series of tracks. Bird was sceptical about the dinosaur's ability to walk on land. *Brontosaurus*, whose name means literally 'thunder lizard', would certainly have made the ground thunder under its feet as it passed, since it weighed as much as half a dozen elephants. The tracks in Texas had been made in stiff mud, which in itself gives us little clue to whether the animals were walking on land or in water. The first clue that there was at least some water came when no tail furrows could be found in the mud; the tail must have been floating behind the wading dinosaur. Had there been no water, it was reasoned, the tail would certainly have dragged in the mud, leaving a groove. Only one such furrow was ever found by Bird, although there is little evidence that it was made by a brontosaur's tail. Were the dinosaurs actually swimming? Bird found one remarkable trail in Bandera County in Texas that provides unequivocal evidence that one dinosaur at least was swimming in deep water. The tracks are those of the front feet only, lightly impressed in the mud; the animal was floating, moving along by kicking the bottom with its front legs. Then there suddenly appears, as if from nowhere, a single hind foot impression, and the animal changes direction. The brontosaur had made a turn in the water by kicking off in a different direction with its back leg. Here was evidence of a

47. Evidence of the most persuasive kind that brontosaurs ventured into water, these tracks of a brontosaur's front feet could only have been made if the animal was buoyed up by water.

totally independent kind that brontosaurs could swim.[32] These discoveries banished as heresy talk of brontosaurs, brachiosaurs or *Diplodocus* being anything but marsh or swamp dwellers, allowed on land only to breed. In many cases the sauropod was relegated to the deep, its head breaking surface every few minutes for air.

Meanwhile, totally unrelated studies were being carried out in another branch of science. German physiologists early this century were attempting to record the maximum pressure a human subject could withstand under water and still be able to breathe. The volunteers were strapped to boards and lowered into swimming baths in a horizontal position, being allowed to breathe through a tube rising to the surface. Robert Stigler, who began this line of research at Vienna

University's Institute of Physiology in 1911, found that subjects (often himself) were physically stopped from breathing if the board was lowered beyond three feet, at which depth the pressure on the lungs became too great. Stigler himself attempted the same stunt at a depth of six feet but the disastrously high blood pressure generated at so great a depth permanently damaged his heart. British physiologists substantiated Stigler's results, adding that subjects could only breathe comfortably to a depth of ten inches, after which breathing became painful. The first (and apparently only) person who caught the relevance of these physiological studies on humans to the seemingly remotely connected dinosaurs was Kenneth Kermack of University College London. In 1951 Kermack took the results of the German physiologists and applied them to the great sauropods with unexpected consequences (unexpected, that is, to an establishment reared on traditional views that had failed to change since Cope's day, although there can be no doubt that Kermack had anticipated the results). Some of the sauropods with long necks raised vertically could reach 40 feet in height and in such a pose they were imagined standing on a lake bottom with their sinuous necks acting as snorkels. To breathe, an animal has to expand its thorax and lungs. Yet the *Diplodocus*'s lungs would have been at a depth of about twenty-six feet (even though *Brachiosaurus* was taller overall, the long front legs raised the thorax nearer the surface, placing the lungs at about the same depth as in *Diplodocus*). At this depth the pressure is about 8 lbs/sq. inch which, if it did not crush the trachea (the tube passing into the lungs), would certainly have prevented the expansion of the lungs. To have been able to breathe at such a pressure the respiratory muscle would have had to have employed tremendous forces, to the tune of many tons' weight![33] The sauropod's problems did not end there. In order to resist collapse and remain functional the blood vessels must have had blood flowing through them at a catastrophically high pressure, the product of the normal blood pressure compounded by the external pressure. Since the air inside the lungs (assuming they were able to resist the crushing effect of the water) was at normal atmospheric pressure, the blood in the pulmonary vessels would have exploded into the lungs. Such a means of relieving the great pressure could not have been resisted by the lung tissues and fatal haemorrhaging would have resulted. (If, as in Stigler's unfortunate case, the sauropod's heart was not irrevocably damaged first.)

Kermack concluded that it was 'extremely unlikely' that sauropods could have achieved what Stigler's subjects could not. Even if they lived in shallow water, the creatures must have raised themselves out of the water to breathe. Kermack thus totally undermined the 'aquatic nostril' hypothesis: since the animal *had* to rise out of the water to breathe, the nostril could hardly have been a deep water or snorkelling specialisation. Writing in 1951 Kermack's views ran directly counter to the prevailing dogma, a dogma that had failed to take into account the animal's physiology. Colbert answered the challenge. 'By analogy with man, Kermack's logic is sound. However, a question can be raised as to the property of basing such an argument upon our knowledge of human limitations in water. After all, man is not an amphibious or aquatic vertebrate.'[34] Colbert then proceeded to use an aquatic animal as *his* model to demonstrate that creatures can survive under water. Whales, he pointed out, can breathe at the surface while the body is inclined at a steep angle leaving the lungs at some depth. Yet, as Colbert himself admits, whales and porpoises have protective armour around the breathing tubes to stop the collapse. Perhaps, suggested Colbert, 'there may have been some modifications in the respiratory apparatus of the sauropod dinosaurs that enabled them also to breathe while the body was completely submerged'.

Kermack retained his doubts. Whales do surface obliquely, he freely admits, but the hiss that can be heard when the blow hole breaks surface is not the lungs expiring air but the great pressure at that depth forcibly expelling it. Whales then rise to the surface before being able to breathe in.[35] A whale is far too specialised a sea creature to be used in comparison to a dinosaur. Structural modifications, such as the loss of walking limbs, are accompanied by great physiological adaptations to the pressure of the water. The sauropod was a four-footed beast and even the most reluctant aquatic protagonist agrees that it could walk on land, even if it did prefer life in ponds. A whale is adapted to one sort of life only, that of the ocean; out of water it suffocates because its lungs cannot expand without the help of the water-provided buoyancy. The dinosaur is a tetrapod specialised for life on land and hardly comparable to the most specialised of marine mammals.

Refutations like Kermack's are not instantaneous in action; science gives up its cherished theories only with the greatest reluctance. A conservative streak prolongs the life of traditional theories even after they become untenable, like the momentum acquired by a car being driven too fast and unable to stop abruptly. Kermack may have, in his own words, thrown a 'grave doubt' upon the then currently accepted mode of life of the sauropods, but a review of the textbooks published two decades later shows how little his views have been heeded. Still we find sauropods standing on lake bottoms and tearing up water plants with their 'feeble' teeth.[36] 'The human mind', wrote Antoine Lavoisier, the French chemist guillotined during the French Revolution, 'gets creased into a way of seeing things.' One might add that the evolving corporate mind suffers no less, since it proceeds by indoctrination from generation to generation. A striking example asserts itself here: since early Victorian times, when Owen pegged his reputation on an aquatic 'whale lizard', sauropods have been beset by this watery image. Had Cuvier never pontificated on the *whale* ribs, nor Owen on the '*whale lizards*', it is conceivable that to this day not a single snorkelling brontosaur would have lived on a lake bottom. But that is counter-factual. Rather, the trend was set by a string of what appears with hindsight misinterpretations, and a 'cetacean' sauropod quickly became entrenched in scientific lore; the image survives to this day. Scientific thought in any age owes more than it realises to its past, and only by understanding the origin of inherited beliefs can we appreciate their continuing worth.

For twenty years Kermack found no support. The change in the climate of opinion only came in 1971 when Robert T. Bakker presented a radical view of the brontosaur in an article in *Nature*[37] (although this was not the first time that he had voiced his ideas). Bakker had departed even further from traditional views, claiming that brontosaurs were plains and forest dwellers, rather like elephants and giraffes. He assembled a great deal of evidence for his critique of existing views, undermining each argument for an aquatic existence in turn. As for the high situated nostril, Bakker argues that this is also found in the ground iguana and desert monitor lizards. Furthermore, aquatic animals have little need of their sense of smell and possess only small nasal openings. In brontosaurs the nasal organ and its opening on the skull roof are large; obviously, they had an acute sense of smell. Severe tooth wear in many brontosaurs suggests that they fed, not on soft water plants, but on coarse terrestrial vegetation. Bakker draws his main strength, however, from the similarity between brontosaur and elephant construction. Aquatic animals like hippos have short legs and shallow rib cages, unlike both brontosaurs and elephants which have long legs and deep rib cages. The hippo, buoyed-up by the water, has only weak legs flexed at the knee and elbow. In contrast, the pillar-like columns in brontosaurs match those in elephants and must have been used to transport the creature overland. If the brontosaur were a water dweller it would not have required such stout limbs. Neither were its feet adapted to standing on mud; hippo toes are splayed-out for support on soft ground, whereas elephants and brontosaurs have short stumpy toes encased in a fleshy pad (sauropod tracks show this well). Overall, it appears that the brontosaurs lived an elephant's life in Mesozoic times. Can we get still closer with our latter-day analogy?

Why did the brontosaurs possess long necks, if these did not function as snorkels? Long limbs and a long neck occur in a land animal like the giraffe where they are used to greatly increase the foraging range, bestowing a great advantage in the search for food. The giraffe is able to browse off the highest leaves where other creatures cannot reach and, undoubtedly, brontosaurs also found it beneficial to have this great vertical range. It enabled them to browse off the tops of the cycads and pines common in their world (terrestrial cycads and conifers were plentiful in the Morrison flora whereas swamp vegetation is hardly known). Aquatic creatures like hippos, on the other hand, are compact and have only short necks and legs because water plants are most abundant in the shallows.

If the sauropods fled to water at the onset of danger, they would undoubtedly have sealed their own fate. The pursuing allosaurs were endowed with long powerful hind limbs and spreading toes, the sort of foot that is well adapted to giving a kicking stroke in water. The sauropods could have escaped the carnivores neither by out-swimming nor out-running them. In all probability, sauropods grouped together to beat off an attacker, perhaps, as Osborn suggested, rearing up and using their front limbs as formidable weapons.

The giraffe model of the brontosaur, a model that Cope was forced to drop for lack of evidence, has been fully reinstated by Bakker. That this conception of the sauropod is winning acceptance is evident from a sympathetic review of Bakker's article in the same issue of *Nature*. *Nature* is one of the foremost scientific journals so any favourable pronouncement from its august columns is of no little importance. Bakker's anatomical attack, claimed the editorial leaving no doubt about the reviewer's conversion, 'completely demolishes' the aquatic hypothesis once and for all.[38] Certainly the combined weight of Kermack's and Bakker's arguments is impressive.

One piece of uncompromising counter-evidence persists, however. How are we to explain Bird's photograph of the isolated impressions of the brontosaur's front feet? This one fact alone demonstrates that brontosaurs must have taken to water at least on isolated occasions. Perhaps, like elephants, they immersed regularly to prevent the skin from cracking and to remove external parasites. Most large mammals visit water holes regularly. Unfortunately, brontosaur tracks, by their very nature, could only be left in the mud around such water holes, so we should *expect* to find evidence of the animals occasionally wallowing like elephants. But this scarcely warrants brontosaurs being declared aquatic, any more than elephants. The debate will assuredly continue.

Bakker concluded his account of the brontosaurs with a note on the rarely mentioned ecological role played by these majestic beasts.

> The impact of brontosaurs on the Mesozoic terrestrial ecology must have been enormous. The herds of different sympatric sauropod species must have opened up thick forest and kept undergrowth from becoming dense, much as elephants do in Africa. Smaller herbivorous dinosaurs adapted to open woodlands and plains must have depended on sauropods for keeping such habitats from being overrun by thick jungle.[39]

The reference to herds is interesting and leads us into another very unreptilian aspect of the sauropods. Hitchcock, it will be remembered, commented on the numerous *parallel* trails of his Connecticut Valley 'birds', and used them as evidence for the presence of flocks; his 'birds' were, of course, dinosaurs. The dinosaur's life was probably far more highly organised than is commonly supposed. Unlike solitary lizards, many dinosaurs were gregarious, travelling in herds like elephants and antelope, constantly on the move to fresh pastures to satisfy their voracious appetites. Bird was the first to realise this as an outcome of his dinosaur-tracking. In his Bandera County quarry in Texas he came upon a dinosaur 'stamping ground,' as he first called it. Upon closer inspection it became apparent that all the brontosaurs, twenty-three individuals in all, had passed in the same direction and, he suspected, were headed towards a common goal. One or two solitary carnivores had passed earlier but the herd definitely crossed as a unit. Elsewhere he found tracks of a big dinosaur moving leisurely, occasionally dragging a foot before picking it up, followed by fifteen more brontosaurs ambling along 'like cows going down a country lane'.[40] If the sauropod was a herding creature, Bird's inability to trace any tail furrows near his fossil footprints may be simply explained. These beasts dare not have dragged their 30 foot tails over the ground for fear of closely following members of the herd trampling on them; instead, tails must have been held stiffly off the ground out of harm's way.

The dinosaur as a herding animal was given support recently by John Ostrom's study of the footprint site at Holyoke in Massachusetts. This site is earlier in geological age and the prints are all three-toed. They can be divided by size into three sorts with the largest most plentiful (it is conceivable that some of the smaller prints were made by juveniles of the same species). One hundred and thirty four individual footprints comprise twenty-eight trails here, and twenty-three of these trails were made by the large-footed beasts. Twenty of these latter trails share the same westerly orientation, whereas the smaller prints seem to be scattered in all directions.

The dinosaurs could not have been browsing since that would have led to erratic prints as the animals foraged to and fro. They seem to have been travelling as a group sharing a common objective. The deviating trails indicate that there was no physical barrier to movement in other directions. Thus the group was not being 'channelled' by, for example, having to keep to a spit of sand running between adjacent lakes. The probability of these trails being orientated in the

48. Both Cope and Osborn toyed with the idea of the colossal sauropod as a terrestrial giraffe-like dinosaur. Only recently, however, has Bakker supplied a plethora of detailed evidence in support of this. *Barosaurus* (above), like other sauropods, used its extraordinary neck to browse off tree-tops where other dinosaurs were unable to reach.

same direction if the individuals were not members of a herd is phenomenal, Ostrom computed the odds at 2×10^{21} : I against! Therefore, it seems safe to conclude that these dinosaurs were not solitary individuals but members of a herd with a goal in view.[41]

One of the most intriguing problems is that of social hierarchy. With the dinosaur's more than limited mental capacity, could it have participated in a structured herd? Was there a leader and a subservient hierarchy of bulls and cows; a harem of cows to a dominant bull; and a crêche for the youngsters? Or was such organisation beyond the dinosaur's ability? Bird imagined young brontosaurs travelling with the herd. Bakker even proposed an internal structure to the group, perhaps with the juveniles protected by an encircling group of adults. Certainly, the youngsters would be very vulnerable. The adult brontosaur towered above all contemporary enemies and, like the elephant, may have been fairly immune from attack, but the young dinosaur shared no such advantage.

Adult sauropod remains are found in abundance but juveniles are very scarce as fossils. On analogy with reptiles, it has been suggested that there was an explosive mortality in early life with only a few individuals lingering on into adulthood.[42] The chance of survival for young turtles, for example, is slim, and to ensure a sufficient number do survive a great many eggs have to be laid. Perhaps young sauropods quickly fell victim to the allosaurs that stalked the herds. It was only when the survivors passed some critical threshold of size that they became immune from attack and their future was assured.

There are compelling reasons, however, why the sauropod did not begin life as an egg left to hatch in the sun. There are physical limits to the size an egg can attain. As the volume of fluid in the egg increases the shell has to become thicker to withstand the pressure, and there comes a point when the shell would be too thick to allow the juvenile to break out at hatching time. The giant sauropods could therefore lay eggs no larger than those of much smaller dinosaurs. This presents two problems. First, if the eggs were laid in the warm sand or soil and left to hatch, how did the youngsters ever find their way to the herd, which would be many weeks away at the time of hatching? Secondly, how could a tiny hatchling no larger than a big lizard ever manage to avoid being trampled underfoot by adults as tall as three-storey buildings? It seems more probable that the female sauropod gave live birth whilst on the move. The youngster would be larger in size and quickly join the herd from which it would derive immediate protection. It also means that less young would need to be born into the herd because of the greater survival potential. With fewer young, and a lifespan of perhaps a century, there would be a slow population turnover and a bias towards really large adults, as we find in the fossil record.

The amount of foodstuff required to sustain a herd must have been prodigious. Since the brontosaur may have had a metabolic rate as high as an elephant's, it probably needed the same amount of food relative to its weight. A three- or four-ton elephant requires I cwt of hay a day,[43] so a thirty- or forty-ton dinosaur could have consumed half a ton of vegetation each day. With each individual member consuming such a quantity of leaves and fruit, the herd would be forced to move on constantly in order to avoid swallowing its immediate environment.

The most serious objection to this is the difficulty in imagining a *Diplodocus* or *Brontosaurus* passing so much food each day through a head no larger than one of its vertebrae. (It is interesting to note that the posterior nasal openings allowed the sauropod to continually pass food through its mouth without interrupting its breathing – the 'aquatic' nostrils, in fact, substituted for a secondary palate.) Like browsing plains mammals today, brontosaurs probably wasted very little of their day in sleep, but spent it constantly foraging for leaves. Lack of chewing teeth is less of a puzzle. *Diplodocus* nipped off the buds, fruit, twigs and leaves with the peg-like mock incisors in the front of its mouth. Lacking the giant molars that an elephant uses to grind its food, *Diplodocus* was forced to swallow the vegetation unchewed. Birds also lack teeth but they achieve the same end using swallowed stones. In the gizzard (a form of pre-stomach) the food is crushed against these stones by the churning action of the gizzard wall, so what passes into the stomach is little different from the well-chewed food of a mammal. In one herbivorous dinosaur – although not a sauropod – found by an American Museum expedition to Mongolia there were one hundred and twelve highly polished stones or gastroliths lying in its stomach region. There is little doubt that these were an integral part of its digestive mechanism.[44] So perhaps the great sauropods similarly swallowed stones to aid digestion; at least, one specimen of *Barosaurus* has been found with seven of these distinctive gastroliths associated with it. Such highly polished pebbles are often found in rock layers that house dinosaur bones. Where dinosaur skeletons are numerous, such as in the Montana Lower Cretaceous beds, there are literally thousands of these stones scattered about, presumably regurgitated by the dinosaurs after becoming too smooth to be of any more use as grinders.

Like the elephant, which depends heavily on its sense of smell, the sauropod used its large nasal cavity to pick up scents carried on the breeze. Wild elephants wave their trunks in the air to catch scents of forest foliage or ripe fruits and the *Diplodocus* presumably used its neck in the same way. To an animal that probably moved from tree to tree stripping it of fruit and leaves such a tracking device must have been indispensable, permitting it the choice of dining localities. The sight of these great herds of sauropods moving through the Jurassic forests, with young the size of fully-grown elephants protected by a ring of weathered old bulls, must have been one of the most awe-inspiring of any period during earth history.

6. A Griffin rescues evolution

Throughout the vast expanse of geological time one can straightway recognise the two prodigious leaps made by the backboned animals. On both occasions what was involved was a conquering of a new and alien world against seemingly insuperable odds. One of these giant strides occurred when the vertebrates first became airborne, thereby opening up a totally new medium to colonisation and an altogether new kind of existence demanding novel and intricate locomotor and sensory structures. How and why terrestrial reptiles first developed feathered wings, powerful flight muscles and the extensive coordinating centres of the brain is still a matter of much dispute. The process of taking to the air must have been a gradual one involving many stages.

But each stage in the development of a wing must have been functional, it must have been selected out of a myriad of random variations to serve some purpose. Natural selection works, after all, by adopting any structure that will give its possessor a positive advantage over those which lack it. But how could there be a piecemeal development of a wing? Surely, such a thing is nonsensical? Only the finished product is functional, or at least, that is how it appeared to some of Darwin's contemporaries a century ago. St George Mivart saw that continued and progressive modification eventually giving rise to a fully integrated structure like a wing was a near impossibility for Darwin's cherished Natural Selection. Instead, Mivart was forced to consider that the wing and all its harmonious accoutrements sprang into being by a sudden jump, as if there were some internal directing mechanism.[1]

But, of course, a bird's wing could not have arisen spontaneously in the one miraculous leap envisaged by Mivart: consequently, the evolution of the wing has remained an enigma. The problem is that *before* a bird can fly it must have feathers, flight muscles and the associated skeletal adaptations like the sternum. But if it must possess these before it can fly why did it develop them at all? Reptiles were not blessed with the foresight to realise that they were going to need feathers if ever they were to get airborne. The answer to the riddle is that each of the components that in totality were to lead to flight must originally have evolved for some purpose other than remaining aloft. Only when all the requisite structures were present could they be switched to a new function. This will be made easier to understand if we turn to the other prodigious leap made by the vertebrates – the colonisation of the land by the fish.

In Devonian times, about 400 million years ago, the most advanced vertebrates were fishes encased in bony armour. Out on land the first insects and archaic plants were quickly establishing themselves. *There* lay the new world to exploit. Although the fish did not abandon the water voluntarily – but only under great pressure – they did leave it, and it is to these pioneers that all subsequent terrestrial vertebrate life owes its existence. What were the prerequisites for land living? Rather than gills any fish venturing on to land would need lungs to breathe atmospheric air. There were two types of fish in existence in the Devonian that already possessed air-breathing lungs: the lungfish (with rather specialised descendants alive today in Australia, South Africa and South America) and the lobe-fin fish (whose only lingering piscine descendant, the celebrated coelacanth – trawled from the ocean bed off East London, South Africa in 1938, 70 million years after its supposed extinction – has converted its lung into a fat-filled bladder). Although rare today, these were the commonest of fishes in the Devonian. They probably lived in swampy lagoons that would occasionally go foul; since the stagnant water lacked sufficient oxygen, air gulping became a necessity if the fish were to remain alive. So bladders that were probably originally gas-filled hydrostatic organs that enabled the fish to remain buoyant now doubled as lungs, helping the gills out in times of oxygen shortage. But to gulp air the fish needed to push themselves off the bottom, and at this point the lobe-fin came into its own. The pectoral fin of these fishes was composed not only of a fleshy lobe but also an internal bony skeleton (that can be matched bone for bone in its proximal region with amphibian and indeed all later vertebrate limb skeletons). When eventually the slowly drying pools were uninhabitable for other fishes, the lung-bearing fishes could remain alive by gulping air. And when the pool dried completely and the fish were stranded out of the buoyant water, the lobe-fin could still breathe by propping its body off the ground and expanding its thorax and lungs to take in air. (Meanwhile, the lungfish had to burrow into the banks to aestivate until the rainy season: the lungfish had met an evolutionary dead-end.) Out of water lungs would be useless and the fish would suffocate without props to raise the thorax off the ground. By employing the natural fore-and-aft motion of the fin the prop could then be used to move the newly instated amphibian in search of more water.

In this way the lobe-fin fishes of the Devonian were already adapted to survival out of water whilst they were still living in it. They had already evolved the necessary equipment. If we now return to the problem of the flying vertebrates, we have to explain why feathers, wings and so on were developed before the proto-birds were able to fly; what purpose had they originally served? A bird cannot launch itself from a tree without feathered wings – it must have evolved these before it climbed the tree. But what was the former function of these beautifully 'preadapted' structures?

In order to solve the problem we must turn to the oldest known bird – and perhaps the most famous (or, in its early days, infamous) fossil of all time – to see how *it* was equipped to cope with flight. Before 1860 there were no feathers known from rocks older than the Tertiary and since, in the words of the proverb, a bird is known by its feathers, it was assumed that no birds had existed as far back as the Mesozoic. Air supremacy, it was reasoned, rested with the pterodactyls, and it was only when these died out at the end of the Cretaceous that the birds were given a chance (some evolutionists even hazarded that pterodactyls evolved into birds). Then in 1860, just a year after Charles Darwin published the *Origin of Species*, Hermann von Meyer found a feather in the pterodactyl-bearing lithographic limestone deposits of Solnhofen in Bavaria. He cautiously described the feather, hardly daring to believe that it actually belonged to a bird – even if it did look avian. He coined the name *Archaeopteryx lithographica* for this 'ancient feather from the lithographic limestone.'[2] Could this solitary feather really have fallen from a bird in late Jurassic times, even before the iguanodons flourished? Von Meyer was at first sceptical of its authenticity; it would have been by no means the first palaeontological hoax. But a careful study of the feather ruled that out. It was certainly genuine.

The following year, 1861, heralded the palaeontological discovery of the century, and the biggest boost that the emergent Darwinism could have hoped to have received. A headless although otherwise nearly complete skeleton of a hybrid bird-reptile was found 60 feet underground in the Ottmann quarry in Solnhofen, Bavaria; a bird because it possessed feathers yet still a reptile because of its clawed fingers and long bony tail. The sensation it caused among the workmen brought it to the attention of the district medical officer Dr Karl Haberlein, who was not slow to realise the worth of such a creature. But he was not thinking of its scientific worth. Acquiring the fossil from the quarryhands in lieu of medical payments, Dr Haberlein put it up for sale. Any museum could buy it for their collection: any museum, that is, that could afford the unheard-of asking price of seven hundred pounds. The doctor invited prospective customers to glimpse the reptile-bird, provided no notes or drawings were made, and then to tender their bids. Eminent collectors came from far and wide to gaze upon the relic. From America, Haberlein entertained Louis Agassiz, the confirmed anti-Darwinian and founder of Harvard University's Museum of Comparative Zoology. Lord Enniskillen, an avid fossil collector whose acquisitions eventually passed to the British Museum, and the Duke of Buckingham made up the British contingent heading for Bavaria. But Haberlein spent a year hedging in an attempt to raise the price, during which time such a mystique had developed around the relic that the Creationists began to accuse their opponents of manufacturing a 'missing link' to rescue their ailing theory. Andreas Wagner, Munich University's anti-Darwinist, working solely from hearsay and secondhand sketches (themselves the product of memory) did his best to rob *Archaeopteryx* of any characters that might make it appear a link and thus ammunition for the Darwinians. A *bird* with a long reptilian tail and clawed fingers would suit the Darwinian purpose well, so Wagner demoted it to a reptile which had acquired feathers independently of birds as 'peculiar adornments.' He rechristened it *Griphosaurus*, the 'enigmatic' or 'griffin lizard.' 'At the first glance of the *Griphosaurus* we might certainly form a notion that we had before us an intermediate creature,' warned Wagner. 'Darwin and his adherents will probably employ the new discovery as an exceedingly welcome occurrence for the justification of their strange views upon the transformation of animals.' In that case, inquired Wagner, where were the intermediates between this link and the true bird or reptile? The first position having fallen to the enemy, the Creationists now demanded intermediates of intermediates, and presumably so on *ad infinitum*. Convinced he had dealt a crushing blow to the strange evolutionary doctrine, Wagner left his adversaries to their 'fantastic dreams'.[3]

Wagner's description of the griffin reached Richard Owen, keen as always to add to the fossil collection of the British Museum (stored at this time in Bloomsbury; Owen was later responsible for rehousing it under a separate roof in South Kensington). In 1862 Owen despatched George Waterhouse, the Keeper of the Geological Department, to Bavaria with £500, the maximum sum the Trustees were prepared to part with. 'The old German doctor is obstinate about his price,' wrote Owen in his diary, 'and Mr Waterhouse has come away empty handed. We ought not to lose the fossil.'[4] They did not. The Trustees eventually paid in full, which took a substantial chunk out of two years' budgets. It was not a bad buy for the most sensational fossil ever discovered. When at last the German museums capitulated, they learned that the bird had flown to England.[5]

And sensational it was in the hands of the English Darwinists. Owen, Huxley's protagonist in the evolutionary arena, may have described it formally, but it was Thomas Henry Huxley, Darwin's champion in battle and the man who ceremoniously dethroned Owen, who saw its worth. A hush of expectancy arose over the audience at the Royal Institution as Huxley rose to deliver one of his popular lectures. How is it, he asked, if one animal type evolves into another, say reptiles to birds, that a great unbridgable gap separates them today? 'We, who believe in evolution, reply that these gaps were once non-existent, that the connecting forms existed in previous epochs of the world's history, but that they have died out.' A mischievious grin broke Huxley's stern exterior. 'Naturally enough, then, we are asked to produce these extinct forms of life.' And Huxley, the magician as always, proceeded to pull *Archaeopteryx* out of his Victorian hat.[6]

No longer on the defensive, evolutionists could emerge from behind Darwin's 'imperfections of the geological record' excuse for the lack of connecting links. Here at last was a creature truly intermediate between birds and reptiles: a reptile with feathers or a bird with clawed fingers and bony tail; however it was viewed, there was no way of denying its transitional nature. In fact, it was exactly the sort of intermediate that Huxley had anticipated before ever it was unearthed. Darwin was delighted.

When the next Solnhofen *Archaeopteryx* came to light – at the Dorr quarry in 1877 – the griffin nature of the creature was endorsed by the undeniable presence of teeth in its jaws (there had actually been suggestions that they were visible in the London specimen as well). Once again the Haberlein family, this time the son, fell upon the scientific relic. The price on its head (this specimen was endowed with one) was a staggering £1,800. The German government was petitioned to stop the precious relic from leaving the country like its predecessor and it was even offered to Kaiser Wilhelm I. His lack of interest brought forth a torrent of abuse from Karl Vogt, the Professor of Zoology at the University of

Geneva. 'His Majesty did not enter into these views. Ah! if, instead of a bird, a petrified cannon or gun had been concerned!'[7] Vogt, a Socialist in the Revolution of 1848, now manned the barricades in the Darwinian struggle. Speaking at the Congress of Swiss Naturalists in 1879, he tried to whip up enthusiasm (and money) to bring this Darwinian trophy to Geneva. But what began as a fund-raising speech to the naturalists degenerated as a result of Vogt's fiery socialist rhetoric into an all-out attack on the Kaiser, Germanic militarism and imperialism, with the Kaiser coming off worst. Vogt himself was offered the fossil at a reduced price, although he too was unable to raise the sum required. But his invective and taunts of cultural barbarism spurred a German industrialist to buy the slab for £1,000 and then sell it to Berlin University for the same.

Even though the debate over evolution has retreated into history and the Southern States, *Archaeopteryx* remains at the centre of controversy. No longer is the concern with *whether* it evolved from reptiles, the focus of attention has now shifted to *which* reptiles were ancestral.

The similarities between *Archaeopteryx* and the small running coelurosaurian dinosaurs are immediately striking. Stripped of its feathers, *Archaeopteryx* has an almost identical skeleton to these bipedal dinosaurs. As Heilmann noted in the 1920s: 'Hollow bones of very light structure, exceedingly long hind limbs with strongly elongate metatarsals and a 'hind-toe', a long narrow hand, a long tail and a long neck, large orbits and ventral ribs – these are bird-features immediately conspicuous' in the coelurosaurs.[8] Nobody could fail to notice them. Yale University's prolific discoverer and describer, Othniel C. Marsh, gave recognition to these similarities by christening one of his new Cretaceous coelurosaur finds *Ornithomimus*, the 'bird mimic', and later Henry Fairfield Osborn named similar specimens *Struthiomimus* or 'ostrich mimic' on account of their uncanny resemblance to the ostrich. But that birds were actually *descended* from these dinosaurs seemed just too hard to believe. Whenever a close affinity was suggested between bird and dinosaur, there followed a torrent of counter-proposals. The debate has continued now for a century; even in the 1870s Samuel Wendell Williston (later to become one of the foremost fossil reptile experts, then a dinosaur collector for Marsh), was campaigning for a dinosaur ancestry in opposition to another of Marsh's prospectors, Professor Benjamin Mudge in Kansas. (The only reason, in fact, that Marsh was able to hold the world record for 'discoveries' of new dinosaur species is that he, or rather his Uncle Peabody's museum, could afford a battery of collectors in the field.)

Even earlier still, Marsh's implacable foe Edward Drinker Cope had been struck by the way these fossil reptiles approached birds in shape. He was especially impressed by *Compsognathus*, a small bipedal coelurosaurian contemporary of *Archaeopteryx*, but was unable to decide whether this little coelurosaur more closely approached the penguin or ostrich.[10] Meanwhile, in London, *Compsognathus* had been noticed by Huxley who was grappling with the same problem, and his solution was announced to the Geological Society of

49. Apart from *Archaeopteryx*, *Compsognathus* (above) was the smallest dinosaur. Could it too have been feathered?

London (of which he was then President) in 1869. He was going to sweep away the 'dinosaur' as an all-encompassing term and bring in a new name – the Ornithoscelida or 'bird leg' in recognition of the bird-like dinosaurian hind leg. Ordinary dinosaurs would remain in one of the subordinate groups, and in the other Huxley set aside *Compsognathus* in order to stress its amazing bird-like qualities.[11] Huxley's bold attempt to overthrow old ideas did not have a long life. One of the personalities in the audience listening to his lecture was the man who was shortly to write the definitive word on dinosaurian classification, Harold Govier Seeley. Nevertheless, it demonstrates the recognition that the avian qualities of the dinosaur were receiving in the early years. And it perhaps excuses Edward Hitchcock for sticking to his belief that the Connecticut Valley tracks were those of giant birds.

The debate raged back and forth over the decades. Although the similarities could not be ignored, most authorities seem to have considered a dinosaur ancestry for the birds heretical to say the least. But in that case the onus was on them to explain away the similarities without involving any sort of direct relationship. In the words of one of the most respected palaeontologists and evolutionary theorists this century, George Gaylord Simpson, 'Almost all the special resemblances of some saurischians to birds, so long noted and so much stressed in the literature, are demonstrably parallelisms and convergences. These cursorial forms developed strikingly bird-like characters here and there in the skeleton and in one genus or another' but they never really approached the avian grade.[12] Convergence was the expedient adopted by the orthodox to avoid the consequences of the remarkably bird-like nature of the coelurosaurians. They protested that the birds and dinosaurs had a distant – probably Triassic – ancestor in common, located within those primitive thecodonts that gave rise to all the archosaurs, whether pterodactyls, birds, crocodiles, dinosaurs and so on. But the small coelurosaurians and the pro-avians had lived similar sorts of lives. Both were runners for both had long legs; both were predators with jaws full of pointed teeth and had grasping hands for catching small mammals and lizards. Living in the same niche, facing the same pressures, eating the same food, natural selection had caused similar looking structures to serve the same functions in both groups. It is a common process as we have already seen: the dolphin shows great convergence to the Mesozoic marine ichthyosaur because it has adopted the same lifestyle in the same sort of environment – both lost their limbs and developed fins for steering and both developed a streamlined body with nearly identical contours. In the same way the skull of the Tasmanian wolf has stumped many a student who believed it to be that of a placental wolf because they are almost indistinguishable at first glance. But the placental wolf is more closely related to the elephant than it is to the marsupial wolf of Tasmania. Living a similar lifestyle has caused them to converge in appearance.

Those who favoured a dinosaurian ancestry were effectively silenced by the publication of Gerhard Heilmann's highly influential book on *The Origin of Birds* which appeared in the mid 1920s (after great difficulty as Heilmann bitterly complained, because none of the English publishers understood its contents or showed the slightest interest in the venture; unless, of course, the author defrayed the total cost![13]) The book has remained a classic ever since. Heilmann seemed on the verge of accepting a dinosaurian ancestry; he listed innumerable likenesses and noted that 'the striking points of similarity between Coelurosaurs and birds pertained to nearly all the parts of the skeleton. From this it would seem a rather obvious conclusion that it is amongst the Coelurosaurs that we are to look for the bird ancestor'. (Compare this statement with the opinion expressed by Simpson above: it demonstrates how experts can view the same material through quite different eyes according, in many cases, to prior theoretical commitments.) But then Heilmann baulks before taking the last heretical step. 'And yet, this would be too rash, for the very fact that the clavicles [collar bones] are wanting would in itself be sufficient to prove that these saurians could not possibly be the ancestors of the birds.' None of the coelurosaurian dinosaurs were known with collar bones, but *Archaeopteryx* possessed them. Since evolution cannot work backwards (a dictum known as Dollo's law) and that if a structure is lost it cannot re-evolve, it is impossible for a group that has lost the collar bones to evolve into a group that possesses them. And that settled the question. One or two intransigents rallied to the side of the dinosaur (notably Percy Lowe in the 1920s and later, who considered *Archaeopteryx* 'a flying dinosaur' with no avian characters whatsoever. On account of its small number of cervical vertebrae he thought it far too specialised to have given rise to modern birds and so he dismissed it as one of Nature's evolutionary blind-alleys.) But these outbreaks were isolated. The search for an ancestor shifted back in time to the Triassic when more primitive creatures still retained collar bones, and the more heavily built pseudosuchians were chosen as the likeliest candidates. Even though the early birds and running dinosaurs looked like one another they had, in

50. Heilmann imagined the avian ancestors as gliding reptiles.

fact, originated independently from the less refined pseudosuchians. Hence the
delicate coelurosaurs, so strikingly similar in nearly all respects to the *Archaeopteryx* that lived alongside them, were ousted because they lacked collar bones.
Any relationship was 'spurious', in the words of Gavin de Beer,[14] and to believe
in it was deemed heresy by the orthodox, a state of affairs that lasted until 1973.

The actual pseudosuchian bird ancestor was assumed to have been the unspecialised bipedal *Ornithosuchus*. This closely approached the most primitive dinosaur, but its name – which means literally 'bird crocodile' – nicely illustrates the supposed common ancestry of the birds, dinosaurs and crocodiles. It may at first sight seem surprising that the closest *living* relative of the bird is the archaic-looking, armour-plated, aquatic crocodile. They appear to have little in common now only because they have radiated into very different niches which require very different modifications. In fact, bird and crocodile skulls share many common characters demonstrating their joint ancestry. But just how close the birds and crocodiles are cannot be agreed upon. Only recently, Alick Walker of Newcastle University came down strongly on the side of a far closer affinity than anyone had previously suspected as a result of his studies on *Sphenosuchus*, which has been variously interpreted as a very primitive Upper Triassic crocodile or a pseudosuchian well on its way to crocodilian status. Nobody is quite sure on which side of the fence it sits. Walker found that many of the bones in *Sphenosuchus'* skull functioned like those in a bird's skull. It even appears that the primitive crocodile had salt-excreting glands in its orbits and so was presumably able to drink sea water and remove the excess salt in the same way that sea birds do today. The inner ear of *Sphenosuchus*, observed Walker, was more like that of a nestling partridge than any present-day crocodile! The precise value of some of the facts marshalled by Walker relating modern birds to ancient crocodiles is a matter of some disagreement, as are his conclusions that the crocodiles evolved from the avian ancestor. 'It thus seems logical to conclude seriously,' says Walker, 'the possibility that crocodiles as a whole may have descended, perhaps as successive "waves", from an unknown stock of late Middle to Upper Triassic reptiles which eventually gave rise to birds, and which may for convenience be called "proavians".'[15] On account of this supposed close affinity (the operative word is 'close' for nobody denies the affinity), Walker is even prepared to accept the possibility that the crocodiles were once tree dwellers.

Walker's letter to *Nature* was a challenge that could not be ignored. It was immediately met by John Ostrom, whose rival letter to *Nature* appeared only a few months later in 1973. Ostrom was in a sound position, having just spent two years studying all the known *Archaeopteryx* specimens (there were four then but one more has subsequently been found) in an attempt to unravel the origin of bird flight. The Jurassic *Archaeopteryx* was obviously far closer to bird ancestry than a partridge and, Ostrom was to show convincingly, the sphenosuchid crocodiles. There was no need to trace the birds all the way back to the distant Triassic crocodiles or thecodonts because the ancestors were already quite apparent in the Jurassic. 'The skeletal anatomy of *Archaeopteryx*,' claimed Ostrom defiantly, 'is almost entirely that of a coelurosaurian dinosaur – not thecodont, not crocodilian, and not avian.'[16] The theory had turned full circle, but now the *evidence* as Ostrom presented it was overwhelmingly in favour of Huxley's dinosaurian ancestry for the birds. So strong are Ostrom's arguments that it is difficult not to call *Archaeopteryx* itself a dinosaur. He listed a plethora of fine details in which the small, bipedal, grasping coelurosaurs resembled the small, bipedal, grasping *Archaeopteryx*. But the reason for the change in the palaeontological climate of opinion rests more with the new discoveries and new interpretations. New dinosaurian finds, especially those in the recently re-explored Gobi Desert, showed that many more features were shared than was previously thought. In particular, the critical lack of a collar bone which caused Heilmann to reject a dinosaur ancestry at the last minute, has been removed as an objection; dinosaurs – including the Mongolian *Velociraptor* collected by the recent Polish–Mongolian Gobi expedition – have now been unearthed with them still present.

But Ostrom's attack was two-pronged. Not only did he show the dinosaurian features of the oldest bird, but he also brought *Archaeopteryx* nearer to the dinosaurs by removing some of its avian features. The typical bird pelvic girdle as a backward-facing pubic process whilst in the dinosaurs it is turned downward. These are diagnostic features; the pelvic girdle has always been pre-eminent in dinosaur classification since Seeley first recognised the differences between the major dinosaurian types in 1887. The Berlin *Archaeopteryx* had always been thought to show this avian back-turning of the pubic bone. Until, that is, Ostrom found evidence that it was probably broken in the fossil and that it could, therefore, have been down-turned in life like any other carnivorous dinosaur.[17] Having cast serious doubt on that, there remains only the furcula, or 'wish-bone', formed by the fusion of the clavicles, to indicate that *Archaeopteryx* was a bird – apart from one rather obvious possession. 'Indeed,' remarked Ostrom, 'if feather impressions had not been preserved all *Archaeopteryx* specimens would have been identified as coelurosaurian dinosaurs. The only reasonable conclusion is that *Archaeopteryx* must have been derived from an early or mid-Jurassic theropod.'

> The most likely origin of so many coelurosaurian features in *Archaeopteryx* is by direct inheritance from a small coelurosaurian ancestor. The additional significance of this phylogeny is that 'dinosaurs' did not become extinct without descendants and I suggest that feathers, as thermal insulators, could be the primary reason for the success of dinosaurian descendants.

The similarities were just too comprehensive, too detailed and too numerous to have been the result of convergence. Convergence as an explanation will stand for just so long before it strains the limits of credulity. Even the Tasmanian wolf skull has small but very important diagnostic features that immediately betray it as marsupial, and hence totally unrelated to our wolf in the Northern hemisphere. *Archaeopteryx* was a small coelurosaur that differed from its contemporaries in its possession of feathers. There should no longer be any need to have recourse to convergence; the overwhelming evidence now demands a direct blood relationship.

Sympathisers heeded the clarion call, seizing upon Ostrom's results as directly relevant to their own work. Suddenly all the pieces fell into place: birds are warm-blooded because they had evolved from already warm-blooded dinosaurs. Endothermy had not evolved independently in these two groups, but only once in the ancestral dinosaur stock. As a corollary, of course, the certainty of warm-bloodedness in birds strengthens the case for the same in dinosaurs. This arrangement is not only more pleasing aesthetically, but it accords with the scientific demand for simplicity; one should not invoke two origins for endothermy when one will satisfy the evidence. Birds inherited their physiological status from dinosaurs.

The implications of this for animal classification are indeed revolutionary. It suggests that we must completely change the way we look at the bird and dinosaur.

If dinosaurs had a single origin, or multiple origins from a restricted group, and if we grant that the ancestral group had undergone a major change in grade from ectothermy to endothermy – a key innovation of such import that it triggered the apparently sudden explosion of dinosaurs into the Triassic world – then our classification should mirror this attainment of the new level of development. It is very difficult arguing *for* the separation of mammals from ancestral reptiles in terms of classification without, at the same time, arguing for the separation of dinosaurs from their reptilian ancestors: the same issues are at stake in both cases. Like mammals, dinosaurs too had made a major break with their ancestors and had risen to a previously unattained peak. Surely, this also justifies them being raised to the status of a Class in their own right, a status that would put them on a par with mammals, reptiles, amphibians and fish? These all represent colossal steps in the history of life: the fish breaking away from the water; the amphibian laying shelled eggs out of water; the reptile first raising its internal temperature to a high stable level without the aid of the sun. Each of these heralds a jump to a new level of organisation.

But what of the birds, why do they not figure in this list? Herein lies a surprise perhaps greater than the creation of the Class Dinosauria. It was first fully appreciated by Robert Bakker and Peter Galton, whose modest four-page letter to *Nature* on March 8, 1974 was titled 'Dinosaur Monophyly and a New Class of Vertebrates'. Modest it may have been in appearance, yet it contains what will probably be the most revolutionary – and controversial – proposal for a major rethink in palaeontology this century. I quote the last paragraph of this key paper disguised among the letters in *Nature*.

Endothermy and high aerobic exercise metabolism are sufficient justification for separating birds into a class distinct from other living sauropsid tetrapods. But endothermy and high exercise metabolism were probably already present in the dinosaur ancestors of birds and are the key features differentiating dinosaurs from crocodiles and other extinct archosaurs. . . . The avian radiation is an aerial exploitation of basic dinosaur physiology and structure, much as the bat radiation is an aerial exploitation of basic, primitive mammal physiology. Bats are not separated into an independent class merely because they fly. We believe that neither flight nor the species diversity of birds merits separation from the dinosaurs on a class level. Among all amniotes, the most profound adaptive shift was from ectothermy to endothermy, which occurred during the origin of mammals and dinosaurs. Therefore we propose the erection of a Class Dinosauria, to include as subclasses the Saurischia, Aves and Ornithischia.

Archosaurs like crocodiles, thecodonts and pterosaurs would stand in relation to the Dinosauria as the mammal-like reptiles stand to mammals. 'This new classification,' they concluded, 'reflects more faithfully the major evolutionary steps. Ectotherms and forms transitional to endotherms are retained in the Reptilia and the two highly successful endothermic groups, mammals and dinosaurs, are given separate class status.'[18]

They are proposing in effect that the birds no longer exist as a Class rivalling the mammals. This privilege now goes to the dinosaurs, of which the birds are living representatives. Birds are just one of three dinosaur groups.

The profound implications of this suggestion could not fail but to be noticed by the popular press in Britain. 'The astonishing claim that birds are dinosaurs', exclaimed *The Sunday Times* in an article entitled "The Flying Dinosaurs", 'cannot be ignored by fellow palaeontologists.'[19]

Bakker and Galton [*The Sunday Times* continued] are proposing a drastic revision in the zoological classification of animals, something almost as controversial as altering scripture. Even those who disagree must take the theory seriously.

But, in spite of the inherently conservative opposition to change that characterises all human activity, science – in contrast to Scripture – *is* a discipline that is constantly evolving, and is therefore amenable to being updated to accord with the latest findings. Science is not the God-given Truth. It only strives towards a greater understanding of the world. There have been many occasions when man has had to radically rethink his world, perhaps none greater than when the earth was removed from the centre of it. Scientific theories only approximate more and more closely towards the truth, they will never reach it. If the evidence is correctly interpreted and the inference follows logically, opposition to the idea will be overcome in time. Alan Charig, the Curator of Fossil Reptiles and Birds at the British Museum, admits that birds may well be descended from dinosaurs, and that dinosaurs may have been warm-blooded, but is unwilling to grant a single origin to all dinosaurs. All three points, argues Charig, must be conceded for the case to be made. As Charig puts it, 'Palaeontologists are not going to start officially calling birds dinosaurs until they are virtually unanimous on these points and we're a very long way from that'.[20] But the argument is equally valid if the Class Dinosauria is accepted as polyphyletic, that is, having multiple origins within the same ancestral group, and so Charig's third point is not really crucial to the case. In fact, none of the existing classes (with the sole exception of the now defunct Aves) is considered monophyletic so this is hardly a prerequisite for the recognition of a Class. The important point in connection with the dinosaurs is that a grade or level of organisation based on endothermy was attained by one or more closely related lines, raising all subsequent lineages out of the cold-blooded condition. Actually, Charig's objection could be met, and our position made more defensible, if the ancestral group – the pseudosuchian thecodonts – were themselves included in the Class Dinosauria. Indeed, there are sound reasons for doing this; it was among early Triassic pseudosuchians, as has already been argued, that the key innovations (erect posture, fast-metabolism and warm-bloodedness) that we associate with the dinosaur radiation first evolved. It would be incongruous to erect a Class Dinosauria and then to leave the pseudosuchians allied to the reptiles. If our new Class is to be meaningful, if it is to mirror the rise in the level of organisation, it must obviously include the pseudosuchians. The Class Dinosauria therefore houses the ornithischians, saurischians, birds and pseudosuchians.

If *Archaeopteryx* was indeed close to the coelurosaur dinosaurs – and Ostrom's work suggests it was very close – and if it was endowed with very little that we would associate with birds but resembled the small dinosaurs in almost all details, what then becomes of its supposed flying ability? After all, it was singularly unlikely that the coelurosaurs could fly and it appears that *Archaeopteryx* was only a feathered coelurosaur *without* the skeletal modifications one associates with flight. It possessed, for example, only minimal flight muscle attachment areas; it totally lacked the huge keel or breastbone that provides the main attachment area for the wing muscles in flying birds.[21] Moreover, the crests on the forearm that receive these muscles are very small in *Archaeopteryx* so there could clearly have been no extensive flying muscles. The shoulder joint is typically dinosaurian and the socket faces downwards, rather than outwards as in birds, and this would have made 'wing' movement more than difficult.[22] Having lowered the wing, it would have been unable to raise it again because the elevator muscles had yet to develop.[23] Instead of possessing less muscles than one of today's birds, *Archaeopteryx* would have required more in order to get off the ground because of its greater weight (its bones were heavy, the weighty skull was full of teeth, and there was a long bony tail).

What bones *Archaeopteryx* did possess were wholly inadequate for a flapping existence. Like the coelurosaurs the elbow, wrist and finger joints were mobile and unfused, whereas in birds the elbow and wrist have only a limited motion and the fingers are fused so the whole forms a solid wing strut to take the strain of flapping. Even supposing *Archaeopteryx* were to have become airborne prematurely, it would not have been able to land safely because it lacked an adequate undercarriage.[24] The pelvic girdle in birds is strengthened and fused to the backbone in the hip region to absorb the impact, while *Archaeopteryx* was fitted with only standard dinosaurian equipment which failed to allow for such contingencies as jumping out of trees. It begins to look as if L. B. Halstead is right in claiming that the oldest bird was no flyer.[25] Flight is the most delicate – not to mention hazardous – form of locomotion yet devised by the backboned animals, yet *Archaeopteryx* had mastered it, or so we are often led to believe, without being incapacitated by its lack of flying muscles!

It is extremely tempting to forget the dinosaurian heritage of this remarkable little feathered fossil housed (rather appropriately) just outside the Dinosaur Gallery in London's Natural History Museum and to imagine it as so many have flapping from tree to tree in search of insects. Recently the problem has been approached from another direction by showing the consequences that would follow if *Archaeopteryx* were to attempt to flap. It seems strange that it should have taken a century before aerodynamic principles were applied to this supposed flyer. In 1970 W. G. Heptonstall of the University of Edinburgh extrapolated some recent findings based on pigeon flight to the pigeon's earliest known ancestor. Of course *Archaeopteryx* – frozen as it is on its slab of rock – is not quite as amenable as a pigeon to the experimental technique, but nonetheless some of the theoretical considerations seem applicable. Since the forearm has to bear the main force of lift it is this that is subject to the greatest degree of stress in flight. Knowing the thickness of the bone and its tensile strength, Heptonstall was able to calculate the maximum weight that the wings could lift and support in the air. He estimated that the *Archaeopteryx* limb-wing could just support the animal gliding but that flapping flight would have imposed too great a stress, and the arm bone would have sheared long before the bird approached hovering.[26] *Archaeopteryx* would have had to have flapped its wings twice as fast as a pigeon to remain aloft, but since it could not muster even one pigeon-power it was destined to remain earthbound.

Those who would like to see *Archaeopteryx* airborne criticise Heptonstall's weight estimate for *Archaeopteryx* of 500 grams, pointing out that modern birds of a similar size are only half this weight.[27] In addition some lift could have been generated by the very long tail, much as it is in Concorde. But it hardly seems reasonable to expect a dinosaur to anticipate the means of weight reduction devised by its descendants during 140 million years of evolution. Birds have had a long time and a lot of practice at losing weight to increase efficiency: fully pneumatic bones, tail-loss and the lightening of the skull by the loss of teeth are just some of the advantages that were not shared by *Archaeopteryx*. Heptonstall's estimate is probably far nearer the mark for a dinosaur that has barely begun to think about flying.

So we are left with the paradoxical situation in which *Archaeopteryx* was unable to fly even though it possessed wings. Why had feathers evolved, if not for flight? The answer to the riddle had long been recognised, even before the realisation that the ancestral dinosaurs were warm-blooded. The clue was provided by modern birds which depend on feathers to keep warm, just as mammals depend on fur. It has also long been argued that the presence of feathers in the *Archaeopteryx* bird-dinosaur was an indication that endothermy was already established.[28] Indeed, looked at logically, it could not have been otherwise. Why would a cold-blooded creature that was dependent upon the external temperature develop feathers? Feathers stop heat loss from a creature actively producing metabolic heat, they do not facilitate heat uptake. As Ostrom observes, feathers would be a positive hindrance to a cold-blooded reptile for 'the insulative properties of feathers also reduce or minimise heat gain from higher

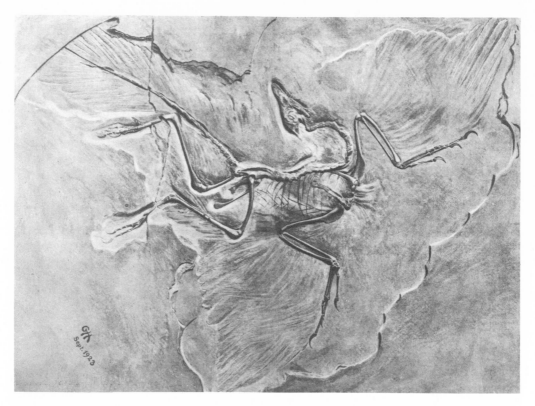

51. *Archaeopteryx* is the smallest known dinosaur, covered in feathers to combat heat loss. This is the Berlin specimen.

ambient temperatures or absorption of solar radiation. That fact clearly establishes that high endogenous body temperatures and effective homoiothermy must have been achieved *before* there was any extensive feather covering. The first feathered creature could not have been an ectotherm! *Archaeopteryx* and its immediate predecessors depended upon internal metabolic heat, not external temperatures. Even more important: this means that endothermy *must* have preceded flight.'[29]

Ostrom's beautiful *reductio ad absurdum* simply stated is this: a cold-blooded creature with feathers would be shielded from the energy source that is vital to its continued existence. Therefore, feathers could not have appeared on a cold-blooded creature. The *Archaeopteryx* bird-dinosaur must have been warm-blooded and, of course, it was because it had evolved from a warm-blooded ancestor.

Feathers provide insulation. Like fur in a mammal they are designed to trap a layer of air next to the skin which can act as a temperature buffer. This air is warmed by the body and cooled by the atmosphere, so the skin itself is never exposed directly to the cold air. Because of this the body is surrounded by a layer of warm air wherever it goes. It means, too, that a warm-blooded animal, which is continually producing heat to remain warm, is not wasting energy through excessive heat loss. The colder the weather, the more the feathers are 'fluffed up'

148

by their superficial muscles to increase the volume of trapped air. It has been argued by those who believed that feathers evolved *for* flying that if insulation was their original function downy feathers would be present in *Archaeopteryx*. But that presupposes that down is a more efficient insulator than the contour feathers that line the body, which is not the case.[30] Down never underlies the contour feathers in modern birds and it is the contour feathers that are controlled by muscles to increase or decrease the insulating air spaces. Although the London *Archaeopteryx* has no visible body contour feathers (these were undoubtably lost after death, such as sometimes happens to a dead gull on the sea shore, leaving only its wing and tail feathers) they do figure on the back, neck, breast and leg of the Berlin specimen,[31] demonstrating that *Archaeopteryx* was totally covered by feathers in life.

Feathers evolved to allow the small warm-blooded dinosaur to cope with heat loss. The operative word here is 'small' for it was the size of the coelurosaurians that explains why feathers appeared in the first place. The dinosaurs, it will be remembered, were restricted to large size by their endothermic physiology. Any attempt to become small would have been thwarted by the excessive heat loss through their naked skin, since a decrease in size is also accompanied by an increase in relative surface area. But *Archaeopteryx* was small, no larger than a crow, and by far the smallest dinosaur known to us (shorter by a foot than *Compsognathus*, its nearest rival). It could only have survived the increased exposure by insulating its skin to prevent heat loss. Birds were the outcome of the dinosaurs' attempt to reduce their size, and feathers the legacy that spelled success.

It is possible that other small Jurassic coelurosaurs were feathered and that the feathers were not preserved because the entombing sediments were not as fine as those at Solnhofen.[32] We have no inkling of the skin-covering of most dinosaurs – only those found as mummies, which are few and restricted to the larger species. *Archaeopteryx* affords crucial evidence that at least some of the smallest dinosaurs were feathered. Once the birds became fully fledged, however, they would have provided stiff competition for any small, feathered non-flyers.

Only after small dinosaurs had acquired them to cope with a large surface area could feathers switch functions and allow the emergent bird to break its ties with *terra firma*. The relevance of the opening remarks concerning 'preadaptation' will now be apparent. Contour feathers evolved as thermal insulators, but they gave the small dinosaurs an aerodynamic bonus that enabled them to conquer the air.

Body contour feathers could only be raised or lowered to increase the insulatory air spaces if they possessed a strong central shaft for the muscles to pull on. And it is only this sort of supported feather that could have trailed behind the arm to form the flying apparatus. Since long trailing feathers would have provided little, if any, insulation, can we assume that possession of such feathers implies an aerial existence? *Archaeopteryx* was adorned with these long wing feathers and if they were not for warmth we must give some account of their function.

It has been commonly accepted that *Archaeopteryx* was a glider and we have seen that its wings were mechanically strong enough to sustain it passively in the air. However, high body weight coupled with the aerodynamically inefficient wings (the claws must have created a great deal of turbulence) resulted in a high sinking speed, which presumably came as quite a shock to its unbraced pelvis. Rather than a graceful glider, the small dinosaur is better imagined plummeting to the earth in an undignified fashion. Nodes on the forearm skeleton of the modern flyer show that the feather shafts are firmly attached to the bone, whereas Ostrom notes that in *Archaeopteryx* these nodes are totally lacking, indicating that its feathers were only superficially implanted in the skin. This undoubtedly made parachuting from *any* height a hazardous occupation.

These speculations presuppose that *Archaeopteryx* was a tree climber in the first place. Yet, paradoxically, trees are not the best place to evolve feathers. To a creature struggling bat-style up a tree an armful of feathers would be a positive hindrance; the more so to *Archaeopteryx* because it was not able to completely retract its wing. If speed was of the essence, either in escaping from ground-living predators or in an attempt to catch winged insects, its chances of survival would have been none too high. Nevertheless, there were positive advantages to moving into the trees at some point, as there is to the colonisation of any vacant niche. The earliest amphibians crawled out of the water (some would say were forced out by the cut-throat competition) and thrived in the uninhabited lake margins, whilst the reptiles made a clean break by removing their eggs (and thus young) from the perils of an aquatic existence by giving them a shell and laying them out on land. It is competition that provides the motive for innovation. The birds reacted in the same way to pressure, by removing their eggs into the trees where they could be incubated in comparative safety. But it seems highly improbable that *Archaeopteryx* had made such a move.

The intricate coordination between eye and hand that is required for tree climbing, let alone the heightening of the senses that must accompany parachuting, would have left some tell-tale signs on the brain of an arboreal *Archaeopteryx*. One has only to think of the effect that tree-climbing linked to manual dexterity has had on the evolution of our ancestor's brain to see the inextricable relationship. Indeed, it appeared at first that the brain of *Archaeopteryx*, re-investigated in 1968 after being pronounced wholly reptilian a generation ago, did show incipient avian characteristics.[33] Harry Jerison found the brain of *Archaeopteryx* to sit midway between that of a modern bird and a *modern* reptile in size, whilst its shape seemed more avian than reptilian, suggesting to Jerison that bird-like reactions were already established. But are modern reptiles really a fair comparison? Not only are they incredibly small-brained, but are (with the partial exception of the crocodiles) only very remotely related to either birds or *Archaeopteryx*. Most dinosaurs were also notoriously small-brained, it is true, but it is becoming increasingly apparent that coelurosaurs and dromaeosaurs, both agile, fleet-footed runners with grasping hands, were endowed with relatively gigantic brains and huge eyes to coordinate their sophisticated reflexes. Since *Archaeopteryx* arose from coelurosaur stock, it would be *expected* to possess a larger than normal brain as standard equipment. No longer can brain size be cited as evidence of *Archaeopteryx*'s tree-climbing habits.

The small bipedal dinosaurs looked so much like *Archaeopteryx* that, as we have seen, the orthodox went to great pains to stress how marvellous convergence is (as indeed it is) that near-identical structures could evolve in distantly related animals: these structures must have had similar functions. But this same orthodoxy had also come to endorse the arboreal hypothesis of bird origins, according to which bird ancestors were tree-climbers that initially took to the air by parachute. As evidence for this they noted the bird-like foot in *Archaeopteryx* with a reversed toe for perching, the grasping hands enabling it to climb, and so on. Such is palaeontological logic that these two mutually inconsistent views could be held side by side for so long. *Archaeopteryx* was, on the one hand, functionally convergent to the running dinosaurs, and on the other, it was a tree-climbing creature! The reversed toe in *Archaeopteryx* was also present in the flesh-eating dinosaurs. The only way out of the paradox if these were convergent or functional equivalents is to assume either that *Tyrannosaurus* perched in trees, or that *Archaeopteryx* was a ground-living predator. Since the first option is rather unlikely, John Ostrom – who was the first to draw out the paradox forcefully – has chosen the second.[34] Those adaptations like the reversed toe and grasping hands that were present in the coelurosaurs, also occurred in the supposed tree-climbing protobirds because they too were bipedal runners.

Baron Nopcsa, that colourful Transylvanian spy, dinosaur expert, and self-styled heir designate to the Albanian throne, had arrived at somewhat similar conclusions in 1907 (although the running hypothesis originated with Samuel Williston in 1879). Nopcsa argued that since the *Archaeopteryx* foot was identical to the dinosaur foot, the earliest bird must also have been a ground-living creature. He believed feathers arose as scale outgrowths from the trailing edge of the arm to increase the speed of the animal as it ran whilst oaring itself along. These feathers would allow the protobird to take large goose-jumps by partially lifting it off the ground. The bird's running ancestry, he thought, was reflected in the surviving ostriches which had retained this primitive condition.[35] Although the 'running' hypothesis is acceptable, this mode of feather acquisition seems unlikely. Since the object of wing development was to increase speed, the bird would have been going the wrong way about it. A glance at the aerodynamic styling of a modern racing car shows that the aerofoils are designed to keep the car *on* the ground, not to raise it off! Loss of traction equals loss of speed.

Ostrom has avoided these difficulties by supposing that the large arm feathers were nothing at all to do with flight or locomotion but were related to predation. *Archaeopteryx* was a fleet-footed predator with good eyesight and powerful hands. It probably caught large insects and the best way to do that was with a feather net: the forearm feathers acted as a snare to trap flying insects. The bird could almost surround the insects in this way before grasping them with its hands. Later it began jumping chicken-style to catch flying insects. In this way

52. *Archaeopteryx* lacked adequate muscles for flight. Ostrom pictures it as a ground-living predator, trapping insects with its splayed wing feathers.

flapping became important in prey catching, and must have been accompanied by more sensitive nervous coordination. The bird-dinosaur, complete with endothermy, feathers, wings and a brain able to coordinate intricate maneouvres, was *completely* 'preadapted' to flight.

Thus were the birds born; not from the pterodactyls as St George Mivart had imagined last century. Mivart was seeking to economise his hypotheses – or, in this case, inventions – by allowing flight to arise just once in the Mesozoic, among the pterodactyl ancestors of flying birds. Like Baron Nopcsa after him, he imagined flightless birds such as ostriches springing directly from dinosaurs independently of flying birds. Pterodactyl ancestry was quickly discredited, but a dinosaur ancestry for flightless birds was an idea that Percy Lowe found attractive in the 1920s. Lowe's beliefs, nurtured over a period lasting a quarter of a century, were first formulated as a reaction to Lamarckism of all things. The views to which Lowe took exception emanated from South Africa where a certain Dr J. E. Duerden, a fellow ostrich expert with the sleuthian title 'Officer-in-Charge, Ostrich Investigations' at Rhodes University, had committed the arch biological heresy of denying a Darwinian explanation to the callouses on the ostrich's breast and rump.[36] Ostriches fall forward when crouching and these thickened pads afford the bird some protection. Duerden believed that these callouses were not genetically coded but arose as a response to friction, rather as in human callouses. But since they first appeared in ostrich embryos (obviously before any friction was applied) the learned doctor concluded that they were characters acquired during the parent's lifetime and passed to the offspring. The parent's behaviour literally modified the structure of its offspring.[37] The classic and most frequently repeated case of supposed Lamarckism centres around the giraffe. This animal, at some point in its history (when it possessed a truncated neck) began stretching to the highest leaves and there found the eating so good that it somehow informed its developing offspring of the obvious advantages of a long neck – and the young giraffe duly appeared with one. It is something of a magical process because there is no known mechanism by which body cells can influence reproductive cells. Both Darwin and Lamarck had believed in such a process, but that is understandable since there was no alternative in their day. But at the turn of the century, neo-Darwinism offered an alternative explanation: mutations acting upon the hereditary material could account for the changes observed in animals, and these random changes (like a slightly longer neck in a giraffe) could then be selected if they were advantageous. The last vestiges of Lamarckism had effectively been banished.

So Duerden's anachronistic beliefs were open to the criticism that there was no mechanism to account for the inheritance of acquired characters. Lowe, however, chose to attack from another standpoint. The callouses, he claimed, were present in the ancestral running dinosaurs, which also spent a great deal of time crouching.[38] Ostriches were an archaic group separated from all other birds, they had never taken to the air; after all, there were no known fossil flying ostriches. Lowe believed that they could be traced back to the Cretaceous diving bird *Hesperornis* (which Marsh himself had once described none too seriously as a 'carnivorous swimming ostrich') which shared the ostrich's flightless predicament. The rhea flapping its wings as it ran across the pampas gave no support to the belief that they were once flyers; Lowe thought that the running dinosaurs also oared themselves along and no one had ever suggested that *dinosaurs* were once flying creatures. Ostriches simply failed to get into the air. In his opinion the ostriches had changed little since they departed the ranks of the bipedal reptiles. He drew strength for this view from the presence of chick-like downy feathers in the adult ostrich, which he assumed to be a primitive feature (as we have seen, they were more probably secondarily derived from shafted contour feathers). The chick-like feathers in the adult ostrich indicated that it had failed to change since Mesozoic times, retaining the primitive condition and never advancing to the flying stage. 'So far as their feather-covering is concerned the Struthiones

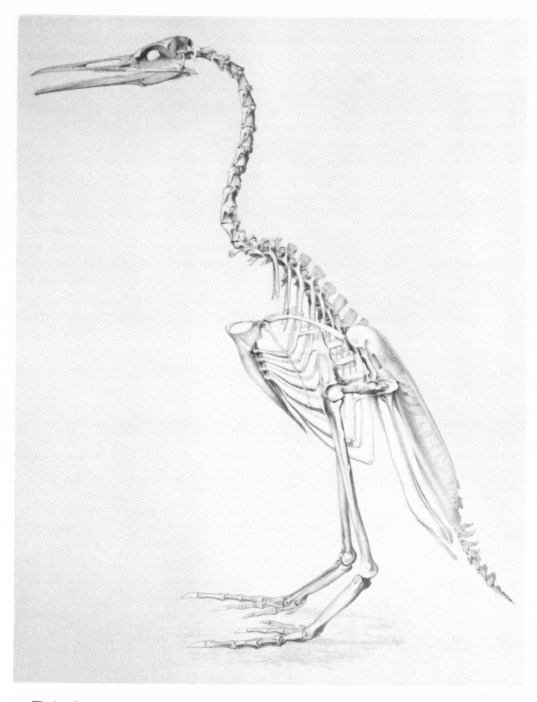

53. The late Cretaceous *Hesperornis* was a marine swimming bird with powerful kicking feet but only vestigial wings. Like *Archaeopteryx* it retained teeth in its jaws. Recent evidence suggests that the American *Hesperornis* migrated north to the Arctic rim to breed.[40]

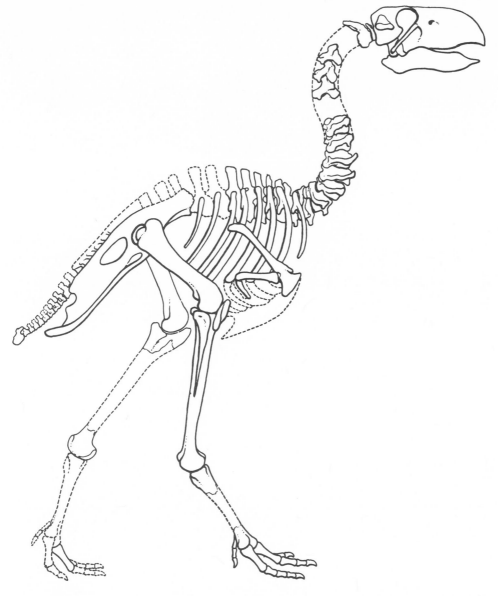

54. Soon after the disappearance of the dinosaurs, powerful ground birds like *Diatryma* (above), found in the Lower Eocene of Wyoming, moved into the vacant niche in an attempt to succeed the dinosaur.

[ostriches] are big, overgrown chicks. They are the 'Peter Pans' of the avian world. They had never grown up.'[39]

It is a pleasant enough concept: dinosaurs degenerating into ostriches. There is a grain of truth in it, of course, but only in as much as all birds sprang from the dinosaurs. The ostrich looks nothing like the bird ancestor, which was fairly small and *Archaeopteryx*-like, if it was not the famous fossil itself. Today, it is uncritically assumed that the earliest birds could fly and that only later did some return to the ground to become runners. The chick-like feathers of the ostrich are due to the adult actually retaining juvenile features: a process known as neoteny. (Humans, too, are neotenous in their retention of an embryonic big head and lack of fur).

Birds, like mammals, had been a subjugated race during the Mesozoic and remained an inconspicuous element of the fauna. They became proficient flyers but were unable to exploit the ground as runners for fear of being hounded by the giant predators. Many birds took refuge in the sea and some, like Marsh's Cretaceous *Hesperornis*, reduced their wings to stubs and developed strong paddling legs. Only with the passing of the dinosaurs at the end of the Cretaceous could birds take to the empty plains. Since the fastest way to progress over flat terrain is to run, we find even early in the Tertiary large and long-legged ground birds. They were attempting to fill the niche vacated by dinosaurs and thus to succeed them as the dominant land animals. Already by the Early Eocene there was one giant, the seven foot tall *Diatryma*, with massive bones and enormous beaked skull, living on the plains of Wyoming, while in England and France at this time ostrich-sized geese roamed over the countryside.[41] During the Eocene birds proliferated to such an extent that it appeared as though the world *would* fall to them. They beat mammals to the large-body niches once monopolised by dinosaurs, and flourished in a variety of large forms. Flightless birds continued to increase in size throughout the Tertiary, developing stout running legs, long necks and good eyesight, and coming to resemble the bipedal dinosaurs. But it was not to last. Mammalian predators took their toll of the large flightless birds and gradually restricted their range. In more recent times man has proved the most destructive of all mammals. Madagascar's primitive fauna was once – only a few thousand years ago – graced by *Aepyornis titan*, the elephant bird. The eggs of this ten-foot bird, themselves about three feet in circumference, are still occasionally found on the Madagascan sea shore. The early Arab traders to these shores, finding these enormous eggs but unable to locate the parents, invented tales of the fabulous Roc, a huge bird strong enough to carry off unwary elephants. The other sanctuary for creatures cut off by time is the Australasian continent, and here the extinct giant moas surpassed even the Madagascan titans. *Dinornis maximus*, whose superlative-laden name translates as 'greatest of the huge birds', reached twelve feet in height. Moas were still lords in their kingdom when the Maoris arrived in New Zealand but their fate was sealed for good in the Maori cooking pot. It is rather an ignominious end to an odyssey begun when the dinosaurs attempted to grow smaller.

7. Phantoms from Hell

When, in 1784, the Bavarian limestone of Eichstätt yielded up its first pterodactyl to the Italian naturalist Cosmo Alassandro Collini, it bequeathed to the scientifically curious all over Europe the ultimate puzzle. Together with the mosasaur, the pterodactyl was a fossil that was to establish in a dramatic way the reality of extinction. Moreover, it made quite clear the great dissimilarity that existed between present forms of life and those of the distant past. Cuvier, who was to assemble the fossil evidence that made extinction plausible, had grave doubts about contemporary interpretations of this strange prehistoric animal. The fossil was housed in the Grand Ducal Museum at Mannheim, where Collini was Curator. Collini himself took it for granted that the creature was a swimmer in ancient seas, but the German doctors who visited Mannheim had other ideas. S. T. Sömmering held that it was intermediate between birds and bats, whilst Lorenz Oken could not decide between a mammal and reptile. Cuvier was not blind to the superficial similarity to many different creatures; he saw in it a resemblance to bats and vampires in general form, its beak was elongated like the bill of a woodcock but armed with teeth like the snout of a crocodile, its vertebrae and legs were like those of a lizard, its clawed fingers like those of a bat, and its body was covered by scales like an iguana's. Cuvier's brilliant comparative anatomical technique had earned him the reputation of the most clear-sighted palaeontologist in Europe. Consequently, it was from him that a solution to the enigma was eagerly awaited. Examining Collini's fossil early in the nineteenth century, paying particular attention to its skull and jaw articulation, Cuvier was able to pronounce with certainty that it was a saurian, a reptile of the ancient world. Its elongated fourth finger bore a wing that enabled it to fly and hence he designated it *Pterodactylus* or 'wing finger'. Although in rank it was a saurian and therefore related to modern reptiles, Cuvier issued a warning: of all the beings in the archaic world this was the most extraordinary and if it were restored to life it would look like nothing else in existence.[1]

It could no longer be doubted that whole races of animals had disappeared from the face of the earth or that primal life was archaic in appearance. Yet so ingrained had become the idea that the pterodactyl had been a swimmer that many quarrelled with Cuvier. Pterodactyls were to prove troublesome in the years ahead. 'In England,' said Harry Govier Seeley later, 'they are classed with the Reptilia, chiefly through the influence of the discourse upon them by Baron Cuvier.'[2] But the German doctors were not to be bulldozed merely on the strength of a Frenchman's reputation. In 1830 Johannes Wagler saw in the

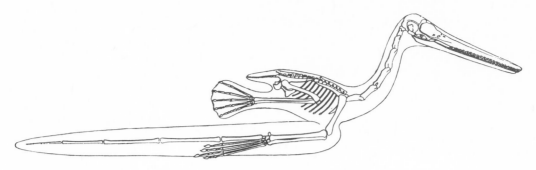

55. Despite Cuvier's plea that the pterodactyl was a winged reptile, many remained faithful to older notions. In 1830 Johannes Wagler endowed it with outsize penguin flippers for swimming.

pterodactyl an affinity to the ichthyosaurs and plesiosaurs that frequented the Mesozoic oceans. Like them it never left the water 'but swam about on the surface like a swan, and sought for its food on the sea-bottom'. He imagined the long arms functioning like penguin or turtle flippers, while the three small clawed fingers were suspected of having been used to grasp 'the females in the generative process'.[3] The following year, 1831, Georg August Goldfuss, Professor of Zoology at Bonn University and a leading German authority on fossil animals, endorsed Cuvier's diagnosis: the pterodactyl was a reptile. But, added Goldfuss, it was a reptile closely approaching the bird in appearance, and probably in consequence of this belief he claimed to see feather-like tufts impressed on the fossil, an embellishment hotly denied by his contemporaries.

Throughout the 1820s many English fossil collectors had guessed that the thin bones in their cabinets were not those of birds but remnants of this strange reptilian creature described by Cuvier. But it was not until December 1828 that a nearly complete skeleton turned up. Nearly all of the early ruling-reptile finds in England were made by amateurs and a great many of them by one in particular, a gifted young lady named Mary Anning. Less well-endowed contemporaries have a penchant for conferring upon the talented some calamity at an early age to account for their abnormal gifts and Mary Anning was no exception. At the tender age of one she and her nurse were struck by lightning, which proved fatal to the nurse, but the infant 'upon being put into warm water, revived; she had sustained no injury. She had been a dull child before, but after this accident became lively and intelligent, and grew up so'.[4] With her new-found abilities she set up shop with her brother in Lyme Regis, Dorset, and there sold the fossils that she collected on her daily strolls along the neighbouring blue Lias cliffs. In 1811, when only 11 years old, she found an ichthyosaur skeleton and in 1824 the first complete plesiosaur. All the geologists of the day befriended her and many went to the then fashionable Lyme Regis for their holidays so they could wander over the cliffs with Mary. When she unearthed the first English pterodactyl in 1828 it was to her frequent visitor the Rev. William Buckland of Oxford University that she presented it. Buckland described the fossil to the Geological Society the following year (although it was not to be seen in print until 1835, a graphic illustration of the more leisurely pace of science in the early nineteenth century; a pace that was to be considerably quickened when Marsh and Cope began jockeying for priority). Even though he had been primed by Cuvier to expect an astonishing sight, Buckland's utter surprise at what he was given led him to reiterate Cuvier's words. It was, he said, 'a monster resembling nothing that has ever been seen or heard-of upon earth, excepting the dragons of romance or heraldry'.[5] Cuvier's anatomical technique, so inspiring to Buckland, had revealed the animal as a flyer. But Buckland was clearly swayed by some of the rival ideas and old confusions reappeared in a new guise. Buckland was perplexed by the short clawed fingers: these must have been powerful paws, he believed, enabling the animal to creep, or climb, or suspend itself from trees. The following year, in his contribution to the *Bridgewater Treatises on the Power Wisdom and Goodness of God as manifested in the Creation*, he exhausted the possibilities by making it a swimmer as well! 'Thus, like Milton's fiend, all qualified for all services and all elements, the creature was a fit companion for the kindred reptiles that swarmed in the seas, or crawled on the shores of a turbulent planet.

> The Fiend,
> O'er bog, or steep, through strait, rough, dense, or rare,
> With head, hands, wings, or feet, pursues his way,
> And swims, or sinks, or wades, or creeps, or flies.
> *Paradise Lost*, Book II

'With flocks of such like creatures flying in the air, and shoals of no less monstrous Ichthyosauri and Plesiosauri in the ocean, and gigantic Crocodiles, and Tortoises crawling on the shores of the primaeval lakes and rivers, air, sea, and land must have been strangely tenanted in the early periods of our infant world.'[6]

The picture Buckland conjured up was actually made flesh by another and rather eccentric Dorset amateur only four years later. Thomas Hawkins, a collector of fossil marine saurians, rendered the contemporary confusion into graphic form and compounded it with theological interpretation that was quite anachronistic even in 1840. The frontispiece to Hawkins' *Book of the Great Sea-Dragons, Ichthyosauri and Plesiosauri, Gedolim Taninim of Moses, Extinct Monsters of the Ancient Earth* had been engraved by John Martin and it portrayed the carnage that resulted when these monstrous beings came together in the pre-Adamite world. John Martin's engravings also adorned Gideon Mantell's *Wonders of Geology*, only here he depicted *Megalosaurus* and *Iguanodon* locked in mortal combat. These engravings received only praise from Hawkins and Mantell although as scientific reconstructions, albeit the earliest of their kind, they look more like caricatures of grotesque apparitions. Othenio Abel, the Professor of Palaeobiology at Vienna University earlier this century and himself something of a reconstruction expert, condemned them for being as outrageous as a Jules Verne fantasy. They remind us, he quipped, of a 'ghostly figure of one of Breughel's hells'.[7]

Hawkins was an indefatigable collector, and in the act of 'disencumbering the old Saurians from their stony shrouds', as Richard Owen put it, ended up

56. *The Great Sea-Dragons, Gedolim Taninim of Moses.* John Martin's interpretation of Thomas Hawkins'
vision of the primeval planet.

carrying away whole quarries. He collected some of the best marine saurians ever found by his method of wholesale demolition and these in time passed to the British Museum where they can still be seen. Luckily, the monsters harmlessly residing in the earth are only 'Effigies of extinguished races', races that 'were as foreign to those of the present day, as are those, probably, of Saturn'.[8] They were an early Creation by Jehovah, who had long since thrown away the blueprint. 'They perpetuate a Design no longer in use,' said Hawkins, so it would be rash to attempt to ally past and present forms. They were worlds, or at least kingdoms, apart. These marine giants were not even of the animal kingdom and Hawkins accordingly created a new kingdom to contain them, the Gedolim Taninim!

Once Cuvier's ideas had won through and the pterodactyl's form of locomotion had been definitely established the confusion should have cleared up. By the time Richard Owen had them constructed at Crystal Palace in 1854 they were recognised by all to be flying creatures. He depicted them perched majestically on a cliff top as if ready to take off, one with wings folded and another with them outstretched. Having reached a consensus that they were flyers a new problem emerged. Since both pterodactyls and birds originated from ancestral archosaurs, and both sported wings, some thought that they may have been directly related. Owen cautiously declined to place pterodactyls on the road to

57. A contemporary illustration of the pterodactyls built by Owen and Hawkins at Crystal Palace. Hawkins imagined the flying reptiles as prototypes of the mythical dragon, and constructed them accordingly.

birds, unlike St George Mivart who, as an ardent evolutionist, desired to make pterodactyls the immediate ancestors of flying birds. Pterodactyls had appeared in far earlier rocks, he noted in evidence, and their long beaks and general outline were reminiscent of birds. A relationship was rendered even more credible when palaeontologists began to obtain fragmentary evidence of the pterodactyl's brain, which had long been suspected of being larger than a living reptile's.

Good fossil skeletons revealing details of the brain are quite rare for pterodactyls, although evidence was pieced together from scraps throughout the nineteenth century. Then in the 1880s the Reverend D. W. Purdon found a promising looking skeleton in the Lower Jurassic Alum Shale at Whitby in Yorkshire which he sent to the Geological Survey in London for examination. At the Survey in 1888 E. T. Newton designated it *Scaphognathus purdoni* to reward the discoverer and, taking advantage of its uncrushed state, he subjected the skeleton to a detailed study.[9] A fracture in the skull allowed a tantalising glimpse of the brain cavity and suggested that a cast of the brain may have been preserved in the rocks for 170 million years. Newton, sacrificing the superficial skull bones, was rewarded by a cranial impression in parts flawless and from this he pieced together a replica of the 'pterosaurian' brain. (By this time, 'pterodactyl' as a collective term was dropping out of use, since it only referred to one of the major groups of winged reptiles; the whole assemblage is now called the Pterosauria.)

The diminutive brain of a modern reptile always fails to completely fill its bony cavity inside the skull. For this reason a cast of the interior of the bony braincase is not really representative of the brain, which is always smaller. In birds and mammals, however, the greatly expanded brain presses on to the walls of the skull, so that an internal cast furnishes an accurate model of the brain in size and shape. Newton, having a good look for the first time at the 'reptilian' pterosaur brain, was amazed. 'The form of the cast exposed leaves no doubt,' he wrote, 'but that it represents the form of the Pterodactyl's brain, just as much as would a cast taken from the skull of a Bird or Mammal; and such casts show the natural external form of the brain even better than the brain itself.'[10] The pterosaur's brain had also completely filled out its braincase.

This was not the only similarity. The brain is characterised by three main regions. The cerebellum, at the rear of the primitive vertebrate brain, deals with movement and balance and is an area little developed in reptiles. The optic lobes of the midbrain coordinate vision and again these are smaller in reptiles and dinosaurs than in mammals and birds. The same is true of the cerebral hemispheres in the forebrain that house the higher coordinating centres (an important exception to this, as we have seen, is found among the agile and alert dromaeosaurid and ostrich dinosaurs). The hindbrain or cerebellum is obviously of tremendous importance in birds for it controls and coordinates the intricate manoeuvres necessary for flight (as well as perching, which requires a far greater sense of balance than standing on four legs). The cerebellum has grown so large in birds that the optic lobes have been forced to one side while the expanded forebrain has grown back to meet it. The enlarged optic lobes have had to expand laterally and ventrally towards the base of the brain.

This is totally unknown in *normal* reptiles, yet Newton encountered exactly the same condition in *Scaphognathus*: the hindbrain had increased in size and had become quite avian in appearance, as had the forebrain, and these two had met and excluded the optic lobes from the dorsal surface of the brain. The optic lobes themselves were large but had been forced to the sides of the brain. The complete reduction of the olfactory bulb demonstates that pterosaurs, like birds, had almost lost their sense of smell and that both depended on acute vision instead. Overall, the similarities between the flying reptile's brain and that of a bird are remarkable. The pterosaur had by far the largest brain (relatively speaking) of any reptile and in terms of intelligence would have made its reptilian cousins seem quite stupid, although among the dinosaurs it could have its match in the dromaeosaurs and ostrich dinosaurs. The pterosaur, supposedly a lowly reptile, had independently evolved a brain nearly as efficient as a bird's. 'Taken by itself,' said Seeley in 1901, 'the avian form of brain in an animal would be as good evidence that its grade of organisation was that of a bird as could be offered.'[11]

The reason for the similarity in brains is easy to understand. The pterosaurs ranged in size from the tiny Late Jurassic 'winged dragon' *Ptenodracon*, which was as small as a sparrow, to the gigantic *Pteranodon* and even larger creatures of late Cretaceous times. Most of the smaller pterosaurs had the short broad wings that characterise the flapping (as opposed to gliding) birds, so it is reasonable to assume that they, too, were flappers. (The long, thin albatross-like wings of later pterosaurs are indicative of gliding habits.) Because they adopted the same sort of aerial existence as birds, pterosaurs would have needed the same coordinating areas in the brain to control balance and permit complex manoeuvres. As in birds, their sight was obviously excellent (this we also know from the large size of the eye sockets) and it is possible that they even had avian colour vision.

Belief that the pterosaur was bird-brained found complete support when another excellent specimen was examined in 1941. Newton's pterosaur had lost some of the cerebellum but the specimen described by the eminent neuro-palaeontologist Tilly Edinger could be completely reconstructed.[12] The new specimen, a tiny *Pterodactylus* no more than a few inches in length and beautifully preserved on its Solnhofen slab, had been presented to the Museum of Comparative Zoology at Harvard University by its founder, Louis Agassiz, in the nineteenth century. A large 'temporal lobe' on the forebrain gave an egg-head appearance to the creature. In life it probably even *looked* intelligent. Edinger also was struck by the way the hind brain and forebrain had grown to meet as in birds. Obviously, she observed, 'this is one of the characters distinguishing all pterosauria from other reptiles'. Any doubts that remained as to the intelligence of the pterosaurs (Newton was reluctant to make too much of this in order not to lend support to the belief that pterosaurs evolved into birds) was quickly dispelled by Edinger. Newton's *Scaphognathus* had an uncommonly long beak which made the brain appear smaller relative to the skull length. In other specimens, however, proportions were like those in song birds. As the pterosaurs evolved their brains became even more bird-like, so that in *Pteranodon* the width

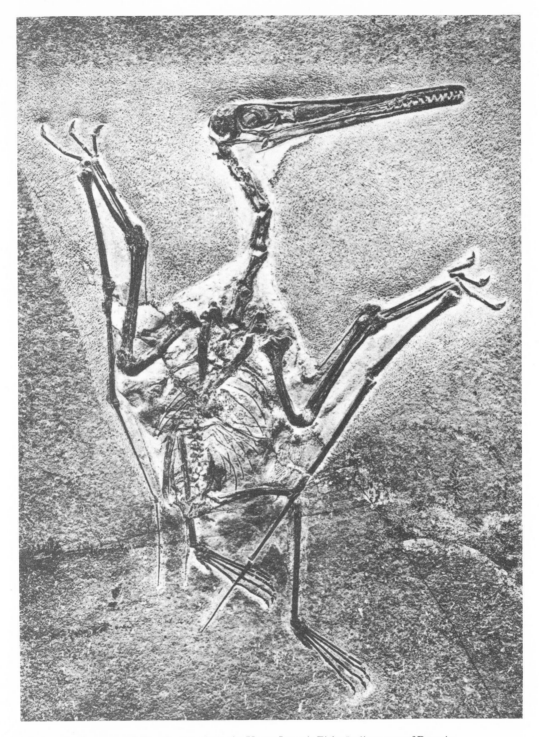

58. A perfectly preserved *Pterodactylus* from the Upper Jurassic Eichstätt limestone of Bavaria.

of the forebrain was greater than its length as it is in birds. The pterosaurs even possessed brain fissures in exactly the same positions as in avian brains, whilst the optic lobes of later pterosaurs like *Pteranodon* were much like those in the most *progressive* birds. In reptiles the hind part of the skull is shaped according to its need for muscle attachment areas, but in pterosaurs, birds and mammals the hind brain had become so large that it produced a cerebellar dome, which itself determines the shape of the rear of the skull. 'As regards length,' concluded Tilly Edinger, 'the pterosaurian cerebella thus stand at the upper limit of reptilian and the lower limit of avian variation' although in appearance they were purely avian. The pterosaur must have been as extraordinary a reptile as a bird or a mammal is! Just *how* extraordinary we have yet to see.

Since an aeroplane's performance matches its controls we can assume that the pterosaurs were very manoeuvrable flyers. The controls were so highly developed for intricate flying that for some time the pterosaurs held their own in competition with the developing birds. The extremely active, flapping existence of birds requires a great deal of energy. The delicate balance, intricate co-ordination and sheer power needed means that the muscles must be constantly supplied with oxygen to rapidly oxidise sugars in order to release a continual stream of energy. These energy producing chemical reactions demand a constant high temperature for optimum fast energy release at all times. It therefore comes as no surprise that the highly energetic warm-blooded bird maintains an average temperature one or two degrees higher than man. Having a fast metabolism, birds are able to remain active for sustained periods. They have gone to extraordinary lengths in developing novel structures to allow the greatest amount of oxygen to be absorbed by the lungs: oxygen required both for the maintenance of a high temperature and for the energy needed for flight.

It is something of a truism that air is drawn into the lungs in a mammal. In birds, however, the air is drawn not only into the lungs but *through* them to extensive air sacs which branch off the lungs and invade and fill out all parts of the body cavity. These provide an enormously expanded surface area for the uptake of oxygen. The branches start as small tubes leading from the lungs and then blow up into air bladders like huge soap bubbles. There are usually ten of these air cavities, which themselves bud off smaller bladders into the arms and legs. The bones in many birds are hollow with small openings to the outside and it is through these that the lung sacs pass to actually line the insides of the bones. The limb bones, backbone and so on are literally lung-filled structures. It is reasonably safe to assume that the smaller pterosaurs were flappers, that is, they launched themselves by beating their wings and were perhaps even able to hover in the air. Presumably, then, the pterosaurs also required a large amount of oxygen for sustained energy production. Since the pterosaurs possessed small openings into their hollow bones, this probably was an identical device to permit the entry of lung sacs. As early as 1837 Hermann von Meyer had used these apertures as evidence that the 'pterodactyl' was a flying animal. The pneumatic or air-filled bones in pterosaurs are very much like those of birds except that the bony casing is even thinner, making the delicate avian bones seem massive by

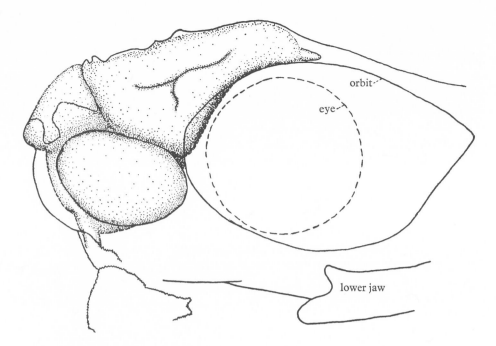

59. The enlarged brain of *Pterodactylus*, pressing against its bony braincase.

60. The egg-head appearance of *Pterodactylus* resulted from its bird-size brain.

comparison. Consequently the bones held a relatively greater volume of air. It is perhaps for this reason that the openings into the bones, similarly placed in both birds and pterosaurs, are always larger in the latter.

Seeley devoted a lifetime to the Mesozoic pterosaurs, and in 1901 summarised his findings in his *Dragons of the Air* which was, apart from an earlier and more specialised work of his, the first book concerned solely with these creatures.[13] In it he was almost loathe to admit that the pterosaurs were reptiles. The brain was bird-like in appearance, proof indeed of the high intelligence required for intricate flapping flight, and the lungs were obviously bird-like and their extensive ramification through the pterosaur's body and bones demonstrated a greed for air that betrayed an active metabolism. Naturally, the conclusion that he drew was that pterosaurs were warm-blooded. Indeed, had it been otherwise, it would be difficult to understand how they could have flown for anything but the briefest period of time without being overcome by exhaustion. A winged (as distinct from gliding) lizard, even if equipped with the necessary musculature and nervous coordination, would be prematurely grounded by its inability to sustain the high energy output required. It is unlikely that the pterosaurs squandered the advantage gained by having avian-style lungs through failing to modify the circulatory system. The lizard's three-chambered heart dilutes the oxygen-bearing arterial blood with oxygen-deficient venous blood before sending it to the body. The crocodile, which split from the ancestral archosaurs at about the same time as the pterosaurs, has a functionally four-chambered heart. This sort of heart was most probably developed in the dinosaurs, also from same stock, and was definitely present in the dinosaurs' avian descendants. As Seeley noted: 'the kind of heart which is always associated with vital structures such as Pterodactyles are inferred to have possessed from the brain mould and the pneumatic foramina in the bones, is the four-chambered heart of the bird and the mammal'.[14] There remained nothing distinctly reptilian, Seeley concluded, about the circulation of the blood in the flying saurians.

Seeley was undoubtedly correct. The extensive ramifications of the lungs throughout the body provided the oxygen that was necessary for sustained flying and endothermy and, equally important, removed the waste carbon dioxide generated by such activity. The efficient four-chambered heart was the pump that ensured the continued supply of oxygen-saturated blood to the tissues. Sustained flying implies a constantly high metabolic rate: the pterosaurs were warm-blooded like birds and mammals.

Since pterosaurs were endotherms and many of them were small (*Ptenodracon* was only sparrow-sized) they must have been protected by a coat of fur or feathers, or perhaps even some form of insulation that has failed to survive to the present day. A warm-blooded creature the size of a bird could never withstand the heat loss if it were stripped of its feathers; there are no naked birds and only one or two examples of small hairless mammals (some burrowers such as the Naked Mole Rat are nearly bald but they are insulated from extremes of cold by their burrows). Goldfuss' claim that tufts of hair could be made out in the fossil pterodactyls was never taken too seriously in Germany. That is not surprising

61. In accepting that pterodactyls were furry, Newman was forced to concede that they could not have been reptiles. So in 1843 he restored them as carnivorous flying marsupials.

since Cuvier had pronounced the beast a reptile, and reptiles are scaly by definition. Until Seeley's lucid study it was hardly to be entertained that a *reptile* could be warm-blooded. Earlier there had been one or two alternative suggestions but the subject was broached with extreme caution and never with any degree of confidence. Goldfuss' ideas made a profound impression upon the English zoologist Edward Newman. But in reviving the notion that pterosaurs were clothed in hair, Newman had also to admit that they could not have been reptiles. Writing in 1843, Newman knew of Buckland's controversial 'opossum' jaw found in the Jurassic Stonesfield Slate thirty years earlier (in fact the first Mesozoic mammal discovered). So he ventured to suggest that the contemporary pterodactyls also were marsupials, only marsupial bats. In the face of ridicule he insisted that anatomists should not adopt a reverential attitude towards authority: Cuvier and Buckland could be wrong. 'Now I believe it within the range of *possibility*', said Newman charily, 'that Cuvier and Buckland should both be in error. I confess that this is highly *improbable*, but I contend that it is *possible*. Regard them as we may, there is still that evidence of humanity about them that induces us to suppose them capable of error.'[15] Newman's heresy was quickly crushed in the ensuing welter of criticism, a fate awaiting any theory based on little more than iconoclastic speculation.

It was only when Seeley began laying the groundwork for the change in opinion regarding the pterosaur's status that the question of insulation again became crucial. If it was warm-blooded, as Seeley suggested, should it not have been insulated? In 1908 Karl Wanderer of Munich University examined the pterosaurs in Dresden Museum to see if *he* could see hair. Goldfuss claimed that his fossil showed impressions of skin with a warty surface, which he interpreted as indicating the presence of hair or feathers in life. Wanderer doubted this. These irregularly scattered 'dimples' were due to nothing more than the flaky nature of the rock.[16] Then in 1927 one of the most respected German palaeontologists, Ferdinand Broili, came out strongly in opposition to Wanderer by insisting that these *were* the impressions of tufts of hair, and to prove the point he compared the warty skin of a bat to the skin impression of the pterosaur. There could be no doubt, he claimed, that pterosaurs possessed a coat of shaggy fur.[17] But his paper was conveniently forgotten and if it were not for some spectacular recent finds it might have sunk completely into obscurity.

In 1970 A. G. Sharov of the Soviet Academy of Sciences, fossil-collecting in the Upper Jurassic lake deposits at Karatay in Kazakhstan, a southern province of the U.S.S.R., recovered a small pterosaur preserved in extraordinary detail. The fossil-bearing rocks in this district, like those of Solnhofen and Eichstädt, are very fine-grained, having formed from the sediments of an ancient lake, and like the Bavarian limestone record the impressions of the skin and other soft parts of the animals that became embedded in them. Sharov's Russian pterosaur came complete with impressions of wing membranes *and* a furry covering. Long thick hair coated the whole body; in life it was probably soft and fleecy since individual hairs and tufts seem to be curved and bent in the fossil. Hair occurred on the fingers and membranes between the toes, although here it was shorter, and it can even be seen on the wing membrane itself. Sharov enigmatically called his beast *Sordes pilosus*, which literally means 'filthy fur'. But he also used the Russian equivalent of *Sordes* and in that language it has the alternative meaning 'devil' or 'evil spirit', which renders a more evocative 'hairy devil'.[18]

This discovery established unequivocally that pterosaurs were warm-blooded. It is easy to say with hindsight that it could not have been otherwise but coming to this realisation was a lengthy and painful process, spanning nearly two centuries. During this time our view of this strange beast had undergone a radical transformation. On the authority of Cuvier the pterodactyl was ordained a reptile, which made hair an impossibility (it is *still* officially classed a reptile in all standard texts). After Seeley's work at the opening of the twentieth century it came more and more to be appreciated that a flying vertebrate could never have become airborne unless it had a fast metabolising, high energy producing physiology, at least on a par with a bat's. This type of physiology goes hand in hand with a high internal temperature, kept stable for optimum biochemical functioning. With this theoretical background Sharov's find came as little surprise; we were primed to accept it. To stop heat loss and energy wastage pterosaurs had evolved fur in parallel fashion to mammals. Like the mammal, the pterosaur probably spent many hours each day preening its fur. In earlier pterosaurs the teeth and claws could have been used as combs; when teeth were lost as a weight-saver in later types and the claws could no longer reach the trunk the beak may have taken over the function.

As the pterosaurs were reasonably intelligent and warm-blooded, what was their mode of reproduction? Did they lay eggs or bear live young? More importantly, was there any post-natal care? Post-natal care is almost unknown in a reptile, which lays its eggs and leaves the young to fend for themselves. If an adult remains near the young, its limited intelligence cannot overcome the temptation to eat them. Such cannibalism is frequently reported among crocodiles.

Since pterosaurs had evolved the lightest flying body of any vertebrate and had taken great pains to reduce weight on a large scale, it is unlikely that the female could have flown carrying offspring that reached a very advanced state of development. If pterosaurs were viviparous the young were probably born in an immature state, an inference also suggested by the very small pelvis of the female. Or, perhaps, like birds they laid eggs. Either way, the hatchling pterosaur, since it was both immature and a warm-blooded creature, would have been dependent upon its parents for warmth, food and later probably flight training. The female pterosaur, being furry and a source of heat, would have brooded the eggs until they hatched (the only parallel case in the reptile world is that of the female Indian python, which wraps around its eggs and heats itself up by sending spasms of muscular contraction down its body). The parent pterosaur would then stay with the young until they could fly.[19] This may mean that pterosaurs paired like birds for a season – or even for life – so the male could supply the brooding female with food stored in his pouch. The living reptile's lowly level of intelligence forbids any such parental care and it is barely able to distinguish its own young from potential food. Pterosaurs, on the other hand, had risen to a new social level, characterised by the temporary family unit. There may even have been a more advanced colonial organisation, with the vulnerable chicks – particularly of the gigantic *Pteranodon* – herded together and protected by one or two guarding adults. (Smaller pterosaurs have been found as thousands of fragments in the Cambridge Greensand in England, suggesting that they may have been gregarious, roosting like bats in colonies. But in this case it is more likely that these bones were washed together long after the death of the widely separated individuals.[20])

Seeley concluded his classic account of the flying dragons, or 'Ornithosaurs' as he called them, with a reaffirmation of his belief that they were something *more* than lowly reptiles: 'these fossils have taught that Ornithosaurs have a community of soft vital organs with Dinosaurs and birds,' he wrote, and they had developed 'the special forms of respiratory organs and brain which lifted them out of association with existing Reptiles'.[21] This was penned at the turn of the century and Seeley, who died only eight years later, was never to know that his ornithosaurs were furry, thus establishing beyond all doubt that they were, as he had always believed, warm-blooded. Their grade of organisation was highly advanced, and although they came from the same ancestral stock as the dinosaurs they had, if anything, progressed even further than their cousins. The dinosaurs were also warm-blooded, it is true, but the pterosaurs had evolved a furry pelt like a mammal, and a brain like a bird, so even if they were not more efficient endotherms, they were certainly more intelligent ones. They had broken the thermal barrier and evolved out of the reptilian grade to just the same extent as birds or mammals. Bakker and Galton claimed, when establishing the Class Dinosauria, that among the higher backboned animals 'the most profound adaptive shift was from ectothermy to endothermy'.[22] This had occurred in mammal-like reptiles and in the dinosaurs' ancestors, giving rise to new levels of organisation and causing sweeping changes in the soft tissues that permitted sustained energy output and higher activity. Bakker and Galton continued: 'endothermy and high exercise metabolism were probably already present in the dinosaur ancestors of the birds and are the key features differentiating dinosaurs from crocodiles and the other extinct archosaurs'. This is not strictly true, as was previously argued, for the very criteria that allow Bakker to pronounce the dinosaurs endotherms are also referrable to the dinosaur's pseudosuchian ancestors; they too were warm-blooded. Finally, this difference may now be extended to include pterosaurs. Endothermy distinguishes dinosaurs from crocodiles, and is a justification for retaining the latter as reptiles and raising the former to a new class. But pterosaurs had also made the break; they had become endotherms and ones, moreover, just as efficient as birds. It seems fairly absurd to leave them behind in the Reptilia whilst raising the dinosaurs out of it. Seeley would not have approved.

If pterosaurs were alive today, we would immediately recognise them as something other than reptilian. Aristotle – the grand collector and classifier of antiquity – would never have grouped a warm-blooded, furry, intelligent animal with the serpents. This tradition of not doing so would have passed down to us in the twentieth century, as it has with birds and mammals. Unfortunately, pterosaurs became extinct at the close of the Cretaceous and it was only from fossil bones that they were reborn in men's minds – bones, moreover, that initially gave little clue to physiology or intelligence. These features have been painstakingly worked out over the decades so that now we have a fairly clear picture of the pterosaur in life, a picture which bears very little resemblance to a reptile.

We recognise that although mammals sprang from reptiles, they are creatures apart. Likewise the pterosaurs, whose most unreptilian physiology is proof that they had long departed from the ranks of the reptiles, should be accorded Class status in recognition of their advanced grade of organisation.

We now have all the evidence before us and can, for the first time, explain the origin and subsequent radiations of the non-mammalian endotherms of the Mesozoic planet, and at the same time understand just why certain groups were able to 'explode' when they did. In short, we can produce a unified theory which relates these radiations of Mesozoic vertebrates to underlying physiological attainments. Pseudosuchians first achieved a warm-blooded physiology in Early Triassic times – 210 million years ago – and themselves developed into paradinosaurs, some later members (for example, *Ornithosuchus*) coming to greatly resemble the bipedal dinosaurian flesh-eaters. In about Mid Triassic times the ornithischian and saurischian dinosaurs made an appearance, exploiting their pseudosuchian physiological and postural inheritance to the full, and quickly assuming a dominant position in the Mesozoic world order. Yet other Mid Triassic bipedal pseudosuchians, with arms freed from the burden of support and locomotion, took to gliding and developed a flap of skin stretching between arms and trunk. We already have evidence of such an intermediate stage, the gliding pseudosuchian *Podopteryx* (literally 'winged foot'), located in Upper Triassic Russian sediments by Sharov.[23] (*Podopteryx* itself was too late in time to have been the actual pterosaur ancestor.) The pseudosuchians were 'preadapted' for flight, in so far as they already posessed a fast-metabolising physiology. Hence they could take to the air as flappers, and as early as the Late Triassic the first pterosaurs appeared in the skies.[24] There was, then, but a single origin for endothermy among Mesozoic 'saurians', which can be traced to the Early Triassic thecodonts, and this triggered two major explosions, the dinosaurian and pterosaurian. These endothermic pseudosuchian descendants realised the potential inherent in their physiological legacy and became some of the most successful terrestrial vertebrates in history. (See the higher vertebrate family tree in Appendix 1.) Substituting a single origin for two or even three independent endothermy origins greatly simplifies our view of vertebrate history, in addition to fulfilling the scientific demand for simplicity, and results in an aesthetically pleasing synthesis, broadly satisfying for its wide scope and great explanatory power (which is, after all, the ultimate goal of scientific explanation).

Despite smaller pterosaurs being flappers, towards the end of the Mesozoic there evolved bizarre gigantic gliders, among them the world's largest flying animals. Before 1840 no species known surpassed a large fruit bat in size. This was not to be wondered at for, said Richard Owen at this time, 'there was *a priori* a physiological improbability that the cold-blooded organisation of a Reptile should . . . be able to raise a larger mass into the air, than could be done by the warm-blooded mammal'.[25] In 1845 the Secretary of the Palaeontographical Society, James Scott Bowerbank, described a new specimen from the chalk of Kent that had a wingspan of at least 8 or 9 feet. Rather than modify his assumption as to cold-bloodedness in pterodactyls, Owen now exclaimed that God was again demonstrating his omnipotence by doing the impossible: this specimen, said Owen, 'shows that the manifestations of Creative power in past time surpass calculations that are founded upon actual nature'.[26] Then in 1847 Bowerbank excavated a pterodactyl that was not only larger than this, but almost twice the size, with a wingspan that reached 15 or 16 feet from tip to tip and which he named *Pterodactylus giganteus*. It was a creature of this size that Owen had reconstructed to full scale at the Crystal Palace in Sydenham.

Following the opening up of the West, the limelight shifted to the New World. In the autumn of 1870 O. C. Marsh (accompanied by Col. William F. Cody, better known as 'Buffalo Bill') led a party from Yale College into the Rocky Mountain region in search of new fossils, and as usual there was a bonanza in store for the eminent palaeontologist. Camping by the North Fork of the Smokey River in Western Kansas, Marsh's team began to explore the Upper Cretaceous Chalk deposits.

Kansas was Cope's territory and the chalk had yielded a rich assortment of spectacular creatures to Marsh's new-found rival, including many mosasaurs and the long-necked marine plesiosaur *Elasmosaurus*. Not that Cope had actually *visited* Kansas. The fossils had, in fact, been shipped back home to him at Philadelphia where he described them. The *Elasmosaurus* was sent in 1868, coinciding in its time of arrival in Philadelphia with the English reconstruction expert Benjamin Waterhouse Hawkins, on the look-out for prehistoric reptiles for his Paleozoic Museum. So it came about that no sooner had Cope described the animal than Hawkins had built a lifesize model in the centre of New York. After the feud between Marsh and Cope (sparked off by Cope not being able to make head nor tail of *Elasmosaurus*) had erupted into the open, the race for discoveries began in earnest. Marsh first crossed the Rocky Mountains in 1868 and, finding the bone pickings in Kansas good, he returned with the Yale party in 1870. Being Marsh, he immediately unearthed three new species of mosasaur, one of which topped thirty feet in length. In the process, he chanced upon some problematical bones; they were long, thin and, upon first impression, resembled the stilt-like leg bones of wading birds. Like avian bones they were light and had thin walls, as well as being pneumatic. The fragmentary bones obviously came from two distinct species. Marsh was troubled because the joint was unlike anything found in a bird and, notwithstanding its great size, could only be compared to the wing-finger joint of the English pterosaurs. But the finger was out of all proportion; although Marsh had only a tiny fragment of one metacarpal, yet it was six and a half inches long. 'This would indicate,' said Marsh prophetically, 'an expanse of wings of not less than twenty feet!'[27] The season's collecting was almost at an end so the excitement had to be contained. Marsh was forced to wait until the following summer before he could confirm his speculations. The next summer, 1871, Marsh's gift for discovery resulted in a find that partially eclipsed even his winged reptiles. Returning to the Kansas Chalk in the Smokey River region, he located the first skeleton of a Cretaceous non-flying bird, *Hesperornis* or the 'Western bird', in the pterosaur-bearing beds.[28] (Cope's sympathisers insist that it was *not* Marsh, but Benjamin Mudge, who actually stumbled on the *Hesperornis* remains and recognised their importance. Mudge was about to ship them to Cope when Marsh appeared on the scene and persuaded Mudge to hand them over.) The giant pterosaurs obviously lived alongside birds of a fairly modern aspect (modern when compared to the archaic *Archaeopteryx*, although, like that dinosaur-bird, *Hesperornis* did bear teeth). Marsh's prophecy of the gigantic size of the flying reptiles was dramatically fulfilled on this trip. Finding new specimens, he was able to add up the lengths of the various isolated wing bones: the metacarpal of the hand, of which he had already found a small piece, was all of sixteen inches long, whilst the first of the four bones of the elongated wing finger alone measured seventeen inches. The upper arm was short at only seven inches but the lower arm bones were longer at thirteen inches. That made a total of over four feet without the three missing bones of the wing finger. 'The above measurements of the wing-bones would indicate for the entire wing a length of at least eight and a half feet, and, for the full expanse of both wings, a distance of eighteen to twenty feet. The present species, therefore, contains some of the largest "flying dragons" yet discovered'.[29] Although it was early days in the race, Marsh could hardly have chanced upon anything more spectacular with which to outstrip his rival. A little higher up in the same beds, he uncovered an even larger pterosaur, whose wing bones were half as broad again as those of the previous specimen, which by now had been christened *Pterodactylus occidentalis*. Marsh, obviously running out of superlatives, saw the new specimen as 'one of the most gigantic of Pterosaurs. It was at least double the bulk of *Pterodactylus occidentalis* and probably measured between the tips of the fully expanded wings nearly twenty-two feet!'

It was certainly a great personal triumph for Marsh. The first American pterosaurs had fallen to him – but only just. Cope, meanwhile, was not to be left out. After Marsh had left Kansas at the end of the second season's dig, Cope immediately moved into the same localities and began frantically excavating to build up his own collection of valuable chalk fossils. (When the *Popular Science Monthly* published an account of Cope's life a decade later, Cope even pre-dated the time of his western explorations to 1870 to make it coincide with Marsh's.) Cope worked quickly as soon as Marsh left in an attempt to get the fossils out and named before Marsh had a chance. Precedence in the naming of the beasts was all important to Cope. On March 1, 1872 he read a paper to the American Philosophical Society describing *his* Kansas pterosaur finds. The two giant winged saurians he named *Ornithochirus* or 'bird feet' in opposition to Marsh's *Pterodactylus*. But he was just too late. As Marsh's mouthpiece – the *American Journal of Science* published in Yale's home town of New Haven – took great delight in pointing out, Cope's paper was distributed in printed form on March 12, whereas Marsh's contribution appeared five days earlier, on March 7. Marsh had retained priority by the skin of his teeth.

The Kansas Chalk proved to be a veritable graveyard of these huge flying saurians and Marsh, followed later by Williston, excavated dozens of fragmentary specimens over the ensuing decades. Marsh renamed his gigantic species

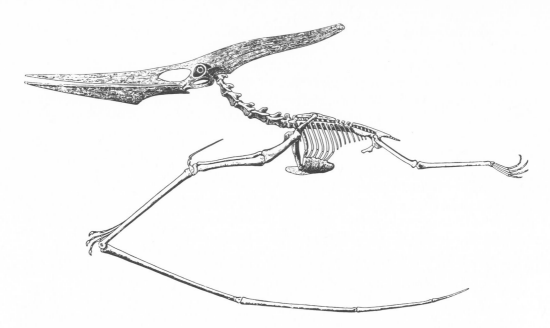

62. The skeleton of the sea-going pterosaur *Pteranodon*. It was constructed like a man-made glider, with the strain of the 12-foot wings taken by a braced, box-like fuselage. The pteroid bone arising from the wrist probably supported an elevator membrane to the neck.

63. It is difficult to understand how *Pteranodon* stood on land. Probably, it propped its body off the ground by resting on its knuckles.

Pteranodon, meaning literally 'winged and toothless'. It was an extraordinary pterosaur in appearance, with jaws drawn out into a long toothless beak and balanced by an elongated bony crest rising from the back of the skull. The creature was literally all wings, which probably exceeded twenty-three feet from tip to tip, whilst its diminutive body was no larger than a turkey. A smaller Kansas contemporary, *Nyctosaurus* or 'night lizard', illustrates the same gross disproportion with a body six inches long supporting eight foot wings.

The long trailing hind limbs in *Pteranodon* were hopelessly inadequate for normal terrestrial movement and it is difficult to visualise the ungainly creatures struggling on the ground at all. Unlike a bat, which is able to fold its wings alongside its body (and even then has difficulty walking on all fours), *Pteranodon*'s body would have been all but enveloped by its extensive wings. Because the wing finger could only be flexed at the knuckle, halfway along its length, the six-foot-long fingers would presumably have pointed skyward as the animal crawled. Any sort of gangling locomotion would have been extremely awkward. Early pterosaurs had a ball and socket shoulder-wing joint but in later ones there was a hinge allowing only up and down motion.[30] 'Perhaps the most feasible method of progression for them when on land,' said Hankin and Watson in 1914, 'is that, having alighted on their feet, they fell over on their stomachs and pushed themselves along, after the manner of penguins, by means of their hind legs, perhaps with an occasional slight lift from the wings for surmounting an obstacle.'[31] Of course, they must have come in to land at some point and therefore must have been able to move on land, although it is difficult to imagine them tobogganing penguin-style, if for no other reason than that they possessed enormous beaks. Perhaps they used their wing fingers for propping the body off the ground. With such clumsy locomotion on land *Pteranodon* would have fallen an easy victim to any flesh-eating dinosaurs in the vicinity. But it seems more probable that the larger pterosaurs were almost perpetual gliders like the albatross, perhaps coming in to land only once a year to breed. The remarkably bat-like rear legs suggest that *Pteranodon* could hang upside down like a bat when on land. Since it was much too large to have been a tree-climber, roosting was thought to have been restricted to rock ledges. However, there were probably no cliffs adjacent to the Kansas inland sea, so it is difficult to accept the popular conception of *Pteranodon* as a cliff dweller.[32] Perhaps it nested on the ground of off-shore islands, where it could exist unmolested.

Like the albatross, *Pteranodon* probably spent most of its adult life soaring. To be able to achieve this and yet grow so large required some fairly drastic weight reduction. It is not usually appreciated just how light this glider with twenty-three-foot wings really was, and without that realisation all sorts of fantastic notions have been devised to allow *Pteranodon* to stay in the air. It had been supposed, for example, that the modern atmosphere was much too thin for these creatures, and that consequently in Cretaceous times the atmosphere must have been twice as dense, allowing *Pteranodon* literally to float on air.[33] *Pteranodon* really had no need of such atmospheric assistance. The animal in life weighed as little as 40 lbs (some estimates range as low as 20 lbs), about 25% of the weight of a man. *Pteranodon* had lost its teeth, tail and some flight musculature, and its rear legs had become spindly. It was, however, in the actual bones that the greatest reduction of weight was achieved. The wing bones, backbone and hind limbs were tubular, like the supporting struts of an aircraft, which allows for strength yet cuts down on weight. In *Pteranodon* these bones, although up to an inch in diameter, were no more than cylindrical air spaces bounded by an outer bony casing no thicker than a piece of card. Barnum Brown of the American Museum reported an armbone fragment of an unknown species of pterosaur from the Upper Cretaceous of Texas in which 'the culmination of the pterosaur . . . the acme of light construction' was achieved.[34] Here, the trend had continued so far that the bone wall of the cylinder was an unbelievable one-fiftieth of an inch thick! Inside the tube bony crosswise struts no thicker than pins helped to strengthen the structure, another 'innovation' in aircraft design anticipated by the Mesozoic pterosaurs.

The combination of great size and negligible weight must necessarily have resulted in some fragility. It is easy to imagine that the paper-thin tubular bones supporting the gigantic wings would have made landing dangerous. How could the creature have alighted without shattering all of its bones? How could it have taken off in the first place? It was obviously unable to flap twelve-foot wings strung between straw-thin tubes. Many larger birds have to achieve a certain speed by running and flapping before they can take off and others have to produce a wing beat speed approaching hovering in order to rise. To achieve hovering with a twenty-three-foot wing spread, *Pteranodon* would have required 220 lbs of flight muscles as efficient as those in humming birds. But it had reduced its musculature to about 8 lbs, so it is inconceivable that *Pteranodon* could have taken off actively.[35]

Pteranodon, then, was not a flapping creature, it had neither the muscles nor the resistance to the resulting stress. Its long, thin albatross-like wings betray it as a glider, the most advanced glider the animal kingdom has produced. With a weight of only 40 lbs the wing loading was only 1 lb per square foot. This gave it a slower sinking speed than even a man-made glider, where the wings have to sustain a weight of at least 4 lbs per square foot.[36] The ratio of wing area to total weight in *Pteranodon* is only surpassed in some of the insects. *Pteranodon* was constructed as a glider, with the breastbone, shoulder girdle and backbone welded into a box-like rigid fuselage, able to absorb the strain from the giant wings. The low weight combined with an enormous wing span meant that *Pteranodon* could glide at ultra-low speeds without fear of stalling. Cherrie Bramwell of Reading University has calculated that it could remain aloft at only 15 m.p.h. So take-off would have been relatively easy. All *Pteranodon* needed was a breeze of 15 m.p.h. when it would face the wind, stretch its wings and be lifted into the air like a piece of paper. No effort at all would have been required. Again, if it was forced to land on the sea, it had only to extend its wings to catch the wind in order to raise itself gently out of the water.

It seems strange that an animal that had gone to such great lengths to reduce its weight to a minimum should have evolved an elongated bony crest on its skull.

The fact that the crest was absent in other pterosaurs suggests that it was not indispensable to the animal. It has been suggested that the crest was chiefly ornamental[37] but it is plainly absurd to imagine so cumbersome an ornament on so specialised a flying creature. An editorial comment in *The Aeronautical Journal* of 1914 by a non-biologist, but obviously a scientist familiar with both the action of evolution and the economy of aeroplane design, put the idea to shame. It 'is obvious that a flying animal that has cut down weight in all other directions is most unlikely to have evolved a large crest either at the expense of, or without effect on, its flying capabilities. That is to say, the crest must have been of aerodynamic advantage to its possessor before its evolution could have progressed nearly as far as it did. It is not difficult to see that such a crest could have been very useful from the point of view of stability. Those who are familiar with the extreme importance of fin disposition on the stability of aeroplanes will not need to have it pointed out that whenever *Pteranodon* pointed its beak downwards it moved a large fin area [the beak itself] backwards and downwards.'[38] But as the beak was lowered the crest was concomitantly raised for compensation: *Pteranodon* was equipped with an automatic trim correction device. It has been demonstrated subsequently that the crest, in fact, *saved* weight. The extra neck muscles needed to keep the beak upright in the face of wind would greatly exceed the weight of the crest itself.[39] Flume tank and wind tunnel experiments with crested and crestless scaled-down models of *Pteranodon*'s head have established this. These experiments also reveal another function for the crest. With the crest removed, the head was constantly twisted to the side by the force of the current and in life the neck muscles would have had to have operated continuously to keep the head facing forwards. With the crest replaced there was full automatic correction when the head was turned obliquely to the direction of flow. In other words, the beak and crest together acted like a weather vane, always facing into the direction of the wind. The crest probably also acted like a rudder in a glider to prevent side to side yawing movements during a turn.[40]

Pterosaur flight has always attracted the aeronautical engineers. The year of Wilbur Wright's death, 1914, saw the most unlikely publishing combine of D. M. S. Watson, the fossil reptile specialist at University College, London, and E. H. Hankin, an aeronautical expert. They studied the intricate series of joints in the shoulder, elbow, wrist and knuckle in the Lower Cretaceous *Ornithodesmus* with a fifteen-foot wingspan. In view of the delicate nature of the bones and the lack of extensive flying muscles in these giant pterosaurs, Watson and Hankin were led to believe 'not that they flew worse than birds, but that they flew more scientifically'.[41] The complicated series of joints gave them an increase in manoeuvrability and perhaps allowed the delicate creature to brake gently whilst soaring.

A similar function has recently been advocated for the puzzling pteroid bone, the small splint of bone on the wing finger for which no use could previously be found. Cherrie Bramwell suggests that it supported a membrane stretched to the neck. On take-off, the bone could be raised and the membrane act as an elevator, as in modern planes. Whilst coming in to land it could be lowered to reduce speed to a minimum without creating turbulence and stalling. *Pteranodon* could land quite safely without damaging its frail bones.

Just how advanced these natural gliders were was discovered when Cherrie Bramwell fed the data for *Pteranodon* into the computer at the College of Aeronautics. The computer was already programmed to test the flight performance of man-made gliders and needed no adjustment for *Pteranodon*.[42] It seems that at speeds above 18 knots (about 20 m.p.h.) *Pteranodon* could swing back its wings into a V (unlike a bat it was able to do this because the wing was supported by one long series of bones which could be flexed at the knuckle). This led to a decrease in the sinking speed and is the principle employed by the swing-wing jet: a principle rediscovered after lying dormant for 70 million years.

The marine sediments that house *Pteranodon* skeletons are also home to a variety of marine reptiles, particularly plesiosaurs and mosasaurs. Some pterosaur-bearing deposits were laid down hundreds of miles from the nearest shore of the sea that invaded Texas and Kansas in Cretaceous times. This suggests that *Pteranodon* could wander great distances from land like the albatross.[43] It obviously lived an idyllic soaring existence, dependent only upon the fish it hunted. Like the pelican, it possessed a throat pouch (which can sometimes be seen in the fossils under ultra-violet light) in which it stored fish. One specimen in the American Museum has the skeletons of two species of fish remaining in its pouch to testify to its dietary preferences.

How the fish were caught presents us with something of a problem. There are a handful of bats which also depend upon fish for food. Like the pterosaurs, they possess huge clawed hind feet which trail behind them as they fly. The mastiff bat (*Noctilio leporinus*), whose habits we know something about, flies out over the water in the evening and gliding down it scoops up the fish in its claws. This is probably the method of the other fishing bats, although their habits are little known. A Californian fishing bat (*Pizonyx vivesi*), which lives by day alongside petrels on rocky crevices, was suspected of being a fisher even before it had been observed at night because of its long rear claws. This was confirmed later by an analysis of its stomach contents, which revealed nothing but fish. But could *Pteranodon* have trailed its claws in the sea to snatch fish? If that was the method and it lived hundreds of miles from land how were fish transfered to the mouth? It is far more likely that *Pteranodon* glided above the wave crests and snapped up surface living fish with its ultra-long beak mounted on a mobile neck whilst remaining airborne. Perhaps it dropped into the sea to take fish. It has lately been suggested that the wing finger was rotated to produce an M-shaped wing plan which would initiate an instantaneous dive.[44] If this was possible at all, it must have been from a fairly low height to lessen the impact of the water on the straw-light bones. An earlier generation, more influenced perhaps by the aeroplane as an instrument of destruction, saw *Pteranodon* in a somewhat different light. Barnum Brown referred to the pterosaurs as the 'airplanes of prehistoric times'. Writing in October 1943, Brown saw a parallel between the terrifying contemporary aerial techniques and the lifestyle of the ancient pterosaurs. The

64. *Pteranodon* was a gentle, intelligent, slow-soaring pterosaur, probably coated in white fur.

'Dive Bomber' *Pteranodon*, as Brown styled it, was the most highly developed flying machine of its day. The crest was 'a sort of rudder which served the animal when, on folded wings, he hurtled through the air for his prey. . . . The neck was moderately long, strong, and flexible. It had a remarkable series of additional articulations unlike anything found in the neckbones of other animals, which gave great thrusting and striking power to the beak.'[45] But Barnum Brown had been unduly influenced by the sinister events of his time. The increasing sophistication of aerial warfare had led him to invest *Pteranodon* with a pugnacity that in life was wanting. It seems unlikely that the frail *Pteranodon* could have withstood the shock of plunging into the water from any height. Unlike gulls, which can furl their wings against the body, *Pteranodon* lacked a ball and socket shoulder joint and so was unable to fold its enormous wings, and left outstretched these could not have withstood the impact. As Watson himself suggested in 1914, at a time when warfare was a little less sophisticated and when aviators knew what fragility meant in the air, the huge pterosaur would probably have disintegrated if it tried to flatten out after a dive![46] Rather than Barnum Brown's demonic dive bomber, *Pteranodon* was a graceful, highly intelligent creature that spent its days placidly soaring over the extensive Cretaceous seas.

Pteranodon spent the daylight hours soaring over the ocean in search of surface living fish. But gliding all day in the tropical sun has its dangers. It is traditionally assumed that these giant pterosaurs were black with leathery wings like those of bats. Yet presenting such an expansive area of wing to the sun could well have led to severe overheating. If the giant outstretched wings were black and leathery they would have absorbed a large measure of the solar radiation falling on them. The extensive blood capillaries permeating the wing would have collected this heat and channelled it back into the small body, which would quickly have cooked. If the animal was as slow a glider as Bramwell suggests the wind would have had little cooling effect. Since we know that Sharov's 'hairy devil' *Sordes pilosus* wore a coat of thick fur and that its wing membranes were also hair-covered, it seems more probable – if at first sight rather startling – that *Pteranodon* was coated in pure white fur. Such a pelt would have reflected the sun and given the creature a still greater resemblance to the albatross. Seagulls are white to afford a measure of camouflage; by presenting less of an obvious outline to surface-living fish they avoid scaring away potential food. This would have been an even greater problem to the fish-hunting *Pteranodon*. Its huge wings would have all but blotted out the sun as it floated low overhead, and black wings would have severely aggravated the situation.

It would be a grave understatement to say that, as a flying creature, *Pteranodon* was large. Indeed, there were sound reasons for believing that it was the largest animal that ever *could* become airborne. With each increase in size, and therefore also weight, a flying animal needs a concomitant increase in power (to beat the wings in a flapper and to hold and manoeuvre them in a glider), but power is supplied by muscles which themselves add still more weight to the structure. The larger a flyer becomes the disproportionately weightier it grows by the addition of its own power supply. There comes a point when the weight is just too great to permit the machine to remain airborne. Calculations bearing on size and power suggested that the maximum weight that a flying vertebrate can attain is about 50 lbs: *Pteranodon* and its slightly larger but lesser known Jordanian ally *Titanopteryx* were therefore thought to be the largest flying animals.[47]

But in 1972 the first of a spectacular series of finds suggested that we must drastically rethink our ideas on the maximum size permissible in flying vertebrates. Although excavations are still in progress, three seasons' digging – from 1972 to 1974 – by Douglas A. Lawson of the University of California has revealed partial skeletons of three ultra-large pterosaurs in the Big Bend National Park in Brewster County, Texas. These skeletons indicate creatures that must have dwarfed even *Pteranodon*. Lawson found the remains of four wings, a long neck, hind legs and toothless jaws in deposits that were non-marine; the ancient entombing sediments are thought to have been made instead by floodplain silting. The immense size of the Big Bend pterosaurs, which have already become known affectionately in the palaeontological world as '747s' or 'Jumbos', may be gauged by setting one of the Texas upper arm bones alongside that of a *Pteranodon*: the 'Jumbo' humerus is fully twice the length of *Pteranodon*'s. Lawson's computer estimated wingspan for this living glider is over *fifty feet*! 'It is no surprise,' says Lawson announcing the animal in *Science* in 1975, 'that the definitive remains of this creature were found in Texas.'[48]

Unlike *Pteranodon*, these creatures were found in rocks that were formed 250 miles *inland* of the Cretaceous coastline. The lack of even lake deposits in the vicinity militates against these particular pterosaurs having been fishers. Lawson suggests that they were carrion feeders, gorging themselves on the rotting mounds of flesh left after the dismembering of a dinosaur carcass. Perhaps, like vultures and condors, these pterosaurs hung in the air over the corpse waiting their turn. Having alighted on the carcass, their toothless beaks would have restricted them to feeding upon the soft, pulpy internal organs. How they could have taken to the air after gorging themselves is something of a puzzle. Wings of such an extraordinary size could not have been flapped when the animal was grounded. Since the pterosaurs were unable to run in order to launch themselves they must have taken off vertically. Pigeons are only able to take-off vertically by reclining their bodies and clapping the wings in front of them; as flappers, the Texas pterosaurs would have needed very tall stilt-like legs to raise the body far enough to allow the 24-foot wings to clear the ground! The main objection, however, still rests in the lack of adequate musculature for such an operation. Is the only solution to suppose that, with wings fully extended and elevators raised, they were lifted passively off the ground by the wind? If Lawson is correct and the Texas pterosaurs were carrion feeders another problem can be envisaged. Dinosaur carcasses imply the presence of dinosaurs. The ungainly Brobdignagian pterosaurs were vulnerable to attack when grounded, so how did they escape the formidable dinosaurs? Left at the mercy of wind currents, take-off would have been a chancy business.[49]

Lawson's exotic pterosaurs raise some intriguing questions. Only continued research will provide the answers.

8. The coming of Armageddon: a cosmic cataclysm?

At the close of the Cretaceous, 70 million years ago, the earth was devastated. Life was ravaged by one of the worst catastrophes ever to have struck the planet: what it was has always been something of an enigma. After its passing, no large land animal was left in existence, no plesiosaurs or mosasaurs remained in the seas nor pterosaurs in the skies, no ammonites survived in the depths of the ocean, nor chalk-forming plankton at the surface. All were annihilated simultaneously. The forms of life persisting – mammals, birds, a few reptiles, land plants, and so on – saw a drastic reduction in numbers: they inherited an earth that would have seemed empty.

This destruction is rendered still more perplexing by the fact that the survivors were the inconspicuous Mesozoic forms. It was just those animals and plants that had flourished in the Mesozoic that suffered the greatest losses. Dinosaurs were among the most successful creatures that ever walked the earth. 'Disregarding the peculiar phenomenon of human evolution,' the Argentinian palaeontologist Oswaldo Reig declared, 'we have to agree that the triumph of the dinosaurs and their relatives has been the major accomplishment in land vertebrate evolution. . . .'[1] Their reign spanned one hundred and fifty million years. If at the Creation Owen's Divine Craftsman had a plan in mind, it was surely the brontosaur in his Mesozoic Eden.

Why, then, the abrupt change in Divine policy about 70 million years ago? What could have precipitated such a crisis? The evidence is scant, and the magnitude of the destruction has generated unprecedented morbid interest with a result that theories are legion. It was argued by a previous generation of palaeontologists on the very grounds of racial longevity that the dinosaurs as a group had reached old age by the end of the Mesozoic. Just as an individual is born and dies, it was claimed, so a race of creatures could become senile after a lengthy period of existence. Evidence for this senility was 'seen' in the apparent over-ossification of the dinosaurian skull: in the development of an abnormal bony frill behind the *Triceratops* skull, the peculiar nasal plume in the duck-billed *Parasaurolophus*, or the grotesque solid-bone domed skull of *Pachycephalosaurus*. Strange forces were also playing upon some of the aged dinosaurian contemporaries. Among the ornate ammonites, some Late Cretaceous members assumed distorted shapes by uncurling and twisting their spirals. These were supposed examples of senile traits, produced in the case of the dinosaurs by a hypothetical malfunction of the pituitary gland. By Late Cretaceous times even the young were born old, appearing from the shell with miniature faces wrinkled with the lines of age. The group, like the individual after its allotted time, was spent and perished of old age. But the analogy is a false one. The process of ageing in an individual can have no parallel in the history of a group. Besides, what does it mean that one group of animals is older than another?[2] Since all types of animals existing at any one time, whether snakes, birds, mammals or dinosaurs, can trace their ancestry back to the primeval ocean they are all of exactly the same age. It is only our artificial construction of a group on the basis of characteristic features – feathered birds, suckling mammals, and so on – which allows us to give it an age. But this procedure unnaturally dissociates the group in question from its ancestors. One could, in fact, retrace all contemporary animals back to the origin of life itself. Furthermore, the 'senile' dinosaurs living at the close of the reign, even those of bizarre or monstrous appearance, were remarkably successful. *Triceratops* roamed in substantial herds, whilst the duck-billed hadrosaurs were perhaps the most common of all late Cretaceous dinosaurs. Many of these so-called 'monstrous' adaptations were of great advantage to their owners. The *Triceratops'* frill or 'neck shield' served as an insertion area for the powerful jaw muscles,[3] and the hadrosaur's crest was an adaptation to increase the sense of smell.

Far from being 'senile', dinosaurs were among the most progressive of terrestrial creatures. During their sojourn as rulers of the earth they produced an array of forms to fill the niches now occupied by mammals and birds as dissimilar as elephants, tigers and ostriches. Dinosaurs were the moderns and constantly moving forward, while the archaic turtles and crocodiles barely advanced at all. In terms of stagnation the turtles were way ahead. Their primitive skull was like that of the ancestor of all reptiles. And while dinosaurs tucked their legs under them as an aid to efficient and speedy locomotion, the tortoises retained the ungainly sprawling gait of the far distant dinosaur ancestor. In all fairness, it should have been the turtle's lot to disappear. Yet survival in changing conditions plays into the hands of the unspecialised; and turtles and crocodiles are still with us, whilst the dinosaur died when our ancestors were still tree-climbing, insect-eating shrews.

The ultra-intelligent 'mimics' unearthed in recent years, dromaeosaurids like *Deinonychus* and *Saurornithoides*, with stereo-vision overseeing grasping fingers and opposable thumbs, and the wide-eyed ostrich-dinosaurs, were a Late Cretaceous phenomenon. These dinosaurs, capable of more skillful behavioural feats than any land animal hitherto, were separated from other dinosaurs by a gulf comparable to that dividing men from cows: the disparity in brain size is staggering. The potential inherent in dromaeosaurs and coelurosaurs for an explosive evolution as the Tertiary dawned cannot be doubted – who knows what new peaks the sophisticated 'bird-mimics' would have attained had they survived into the 'Age of Mammals'. Yet, apparently, not a single breeding population of these beautiful, alert dinosaurs outlived the comparatively cumbersome and dim-witted giants. This is as much a problem as the global nature of the extinctions. The agent we seek was no respecter of intelligence.

Dinosaurs were by no means a static group. Just as they had evolved from Triassic pseudosuchians, so there was a constant turnover within the group. To give an example, the Lance formation of the latest Cretaceous is characterised by the three-horned *Triceratops*, which cannot be found in older rocks; the small, primitive *Leptoceratops* with only an incipient frill was a contemporary, while *Anchiceratops* is to be found in the slightly older Lower Edmonton formation of the Canadian badlands surrounding the Red Deer River; and *Chasmosaurus* and the single-horned *Monoclonius* are ceratopsians of the still older Belly River formation.[4] Each of these Upper Cretaceous formations had its characteristic representatives, ones which usually cannot be found elsewhere. An even more 'revolutionary turnover' is discernible among contemporary ammonites. Here there was a constant and almost complete change of whole communities requiring only a million years for the revolution.[5] It has often been stated that such replacements are the rule in nature and that at the end of the Mesozoic there were simply no successors forthcoming. But why were there no successors? And why should lifeforms as diverse as dinosaurs and ammonites, plesiosaurs and many types of plankton, pterosaurs and numerous land plants all be eliminated simultaneously in what was one of the most all-embracing mass extinctions in geological time?

Dinosaurian extinction must be seen in this broader context. Too many 'solutions' to the riddle of extinction have focused solely on dinosaurs while neglecting the demise of contemporary animals and plants. And all too often the 'solution' put forward to explain the loss of the largest beasts is in no way applicable to the smallest. It is only by taking a broader view of Cretaceous happenings that we can hope to piece together the single cause of *all* the synchronous extinctions at this time.

If we look at dinosaurs *in vacuo* the result is a bewildering array of theories to explain extinction, some of which receive more attention than others. Especially favoured is the view that changing vegetation adversely affected the colossal herbivores. This notion has been a popular one, if for no other reason than that the flowering plants came into existence in the early Cretaceous, about 120 million years ago, and the extinction of the dinosaurs 'followed' at the end of the Cretaceous. And yet the plant-eating hadrosaurs and ceratopsians flourished *in the later Cretaceous*, which strongly militates against their having succumbed to some sort of chemical warfare waged by the flowering plants.

Tony Swain, a biochemical systematist at the Royal Botanic Gardens in Kew is the chief proponent of the theory of 'chemical aggression by plants'. Until the Cretaceous the dominant floral elements were seed-bearing non-flowering plants like ferns, conifers and cycads. Swain analysed the complex chemicals synthesised by plants during their history, in particular those that had acted as feeding deterrents to large destructive herbivores. The ferns and conifers of the early Mesozoic employed condensed tannins to this end, but these were plainly in no way deleterious on a large scale to the pre-Cretaceous dinosaurs. About 120 million years ago the earliest flowering plants replaced these with a new sort of poison, potent alkaloids. Alkaloids are bitter to taste, some like strychnine are highly toxic, others such as morphine produce psychic effects, so could these have adversely affected the dinosaurs? The function of these repellent chemicals was to deter prospective plant-eaters, not to kill them. Swain studied the reactions of tortoises and mammals fed the poisons, finding that for tannins the limit of toleration was similar in both: a certain concentration will kill any desire to eat more. But for alkaloids a serious discrepancy enters. Tortoises need a concentration 40 times greater than that fed to mammals before they are sensitised and cut down their intake. Swain extrapolates these findings to dinosaurs and suggests that – *since both tortoises and dinosaurs are reptiles* – the dinosaurs were 'unable to detect [poisonous] compounds in low enough concentration to be harmless, [and] may well have eaten enough to suffer severe physiological disturbance, and even death'.[6] Because of their voracious appetites, the quantities of alkaloids consumed by dinosaurs would have been considerable. Unaware of their plight, they poisoned themselves to death – and extinction.

There are serious flaws in Swain's arguments; some of his assumptions are unjustifiable; and a serious discrepancy in his timing of the events in the drama make his attempted linking suspect. If the toxic alkaloids appeared 120 million years ago with the first flowering plants, why did the dinosaurs 'linger' on for fifty million years before being struck down? And why, when that time was at last reached, was the event sudden? Even Swain concedes that in its accelerated final phase the total annihilation of the entire dinosaurian stock took only five million years, which, in geological time, is rapid. How are we to explain the simultaneous disappearance of fish-eating pterosaurs from the skies, plesiosaurs and mosasaurs from the seas, and the smaller mammal-eating dinosaurs, as well as a host of other extinctions? But the most serious objection that can be levelled against the theory is that it was only after the introduction of the flowering plants – perhaps as much as 30 million years after – that dinosaurs hit their peak of diversity, a time that saw gigantic herds of plant-eating duck-billed hadrosaurs and horned ceratopsians, as well as the tank-like ankylosaurs. Many of these were in all probability called into existence by the availability of this new plant material. It can have been no coincidence that so many plant-feeders should follow on the appearance of the flowering plants. Colbert states in *The Age of Reptiles*:

The ornithopod dinosaurs, which had enjoyed a moderate degree of evolutionary variety in the late Jurassic and early Cretaceous times as camptosaurs and iguanodonts, increased fivefold in late Cretaceous times, largely as a result of the diverse radiation of the duck-billed dinosaurs. Moreover, two new large groups, or suborders, of plant-eating dinosaurs arose during the Cretaceous period, the armoured dinosaurs or ankylosaurs, and the horned dinosaurs or ceratopsians. Both of these dinosaurian groups became varied and numerous during the final stage of Cretaceous history, thus adding to the array of plant-eating dinosaurs throughout the world. And as might be expected, with so many herbivores inhabiting the lands, there was an increase among the carnivores, the meat-eaters that preyed upon the inoffensive plant-consumers.[7]

But in the hadrosaur's case, as Ostrom has demonstrated, the massive battery of teeth and complex grinding mechanism were designed to deal with abrasive material like conifer needles, not soft flowering plants. Flowering plants could not have been the hadrosaur's poison. This being the case, Swain's arguments have no bearing on the group and they should have survived the Cretaceous.

The tortoise is not a good model – even if it is one of the few plant-eating reptiles on the earth today – for the simple reason that it is as distantly related to the dinosaur as it is to the mammal. In addition, there is every reason to suppose that the dinosaur's physiology was more akin to a mammal's or bird's than to an archaic reptile's.[8]

One of the arguments brought out in support of alkaloid poisoning centres on the contorted pose assumed by many of the delicate, long-necked dinosaurs (especially coelurosaurs) after death. Strychnine poisoning does indeed induce an opisthotonic condition, in which spasms of muscular contraction generated by disruption of the nervous system cause a tightening of the back and neck muscles. This recurves the spine and in long-necked animals throws the head back over the body. Many of the smaller, more agile dinosaurs are found fossilised in this condition, which has led to repeated charges of poisoning with death following in agonised convulsions. This 'pathological' condition, however, is found in all types of dinosaurs throughout their history; so it not only occurred in the Cretaceous long-necked *Struthiomimus*, but also in the Jurassic carnivore *Compsognathus*, which lived before the advent of flowering plants. Many mammals and birds are encountered in a similar contorted pose.[9] There is a far simpler explanation at hand. Skeletal flexure after death is also brought about by the drying out of the long neck ligaments, causing them to shrink and pull back the head, simulating an opisthotonic death. The ubiquity of the finds showing this makes a pathological explanation implausible, especially when the normal process of post-mortem dehydration will produce an identical condition.

There is no denying that the most prominent feature of the Cretaceous-Tertiary boundary is the loss of the dinosaurs; indeed, this event often defines the boundary itself. The sudden and obvious demise of the most dramatic element of the fauna has resulted in a certain short-sightedness in looking for its cause. We tend to overlook the fact that in the depths of the ocean less spectacular organisms were also being destroyed. If we are to see dinosaurian extinction in context – as part of a world-wide, cross-faunal, trans-environmental catastrophe – it could be instructive to turn from the most conspicuous member of the Cretaceous landscape and focus instead on the less dramatic members of its seascape.

The problem of the boundary separating the end of the Mesozoic from the earliest Tertiary was the subject of the Twenty-First Session of the International Geological Congress held in 1960. The reports submitted by the delegates laid great emphasis on the revolution in the microscopic world of the chalk-forming coccoliths and foraminifera at this time.[10] These are single-celled planktonic organisms, and being encased in hard shells they can be traced through the rocks, despite their diminutive size. They give us an insight into marine conditions

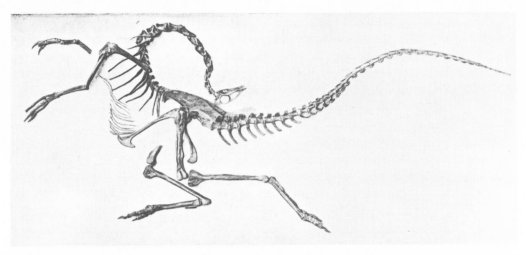

65. The ostrich dinosaur *Struthiomimus*, like so many gracile, long-necked coelurosaurs, is often found preserved in a contorted pose. This is due to the tightening of the neck ligaments after death.

when the last dinosaurs were abroad on the continents. The delegates to the Congress were unanimous: throughout the transition from the latest Cretaceous beds (termed Maestrichtian) to the lowermost Tertiary beds (Danian) there is in each country a marked microfaunal switch. The end of the Cretaceous witnessed the ousting of existing foraminiferans and chalk-forming plankton, while at the base of the Tertiary blossomed foraminiferans of a totally different type. The two planktonic assemblages were absolutely distinct, even though to the in-experienced eye there may be little to distinguish them. The magnitude and abruptness of the switch was evidence that a revolution had taken place in the sea. Obviously these miniscule organisms, drifting at the mercy of the currents, had suffered no less than the dinosaurs. More than one delegate saw an answer to the riddle in a sudden drop in temperature. William Hay, whose special area of study was Mexico, summed up the prevailing opinion:

> The planktonic fauna of the [earliest Tertiary] is composed of relatively few species – all singularly small in size. It resembles the modern boreal [temperate] planktonic fauna and the suggestion is that the ocean waters were much cooler during the Danian than immediately before or after.
>
> The work done in Mexico supports the view that one of the major catastrophes of earth history took place at the end of the Maestrichtian. At the close of the Maestrichtian there was a major regression of the sea documented not only in Mexico, but throughout much of the Tethys region and elsewhere. At the same time the ocean waters were cooled drastically. The great climatic changes which must have taken place probably were the cause of the extinction of thermophilic ['heat-loving] biologic groups so characteristic of the late Mesozoic. The cold water Danian fauna represents the start of a new evolutionary sequence.[11]

In other words, even though they were cosily insulated in the ocean, these tiny planktonic creatures were feeling the effects of a deteriorating climate. Those that were resistant to cooler water were few and resembled the plankton of

189

modern temperate zones. Those that were not perished with the dinosaurs.

The plants were faring no better on land; they were profoundly affected as the Cretaceous passed into the Tertiary. Tropical plants were evicted by their hardier relatives sweeping down from the northerly temperate regions. Since pollen grains are the most resilient parts of plants they are often preserved in the rocks, leaving a record of the plant life of each period. In eastern Montana fossil pollen analysis reveals that half of all late Mesozoic plants were thermophilic flowering plants; in the earliest Tertiary they accounted for less than a third. The evergreen conifers, whose spiky leaves enable them to tolerate colder conditions, had increased their hold over the land as a result of the abrupt change in the world's climate.[12] It was not only flowering plants that conifers were replacing. Typical Mesozoic plants like cycads – for so long the staple fare of many herbivorous dinosaurs – favoured regions with high stable temperatures and for this reason they are today confined to such equable areas as New Zealand, South America and South Africa. As the temperature dropped at the end of the Mesozoic these too lost their hold on the more northerly regions to such conifers as *Metasequoia*.[13] The implications of these studies on plankton and fossil land plants are clear: the earth was plunged into a cold period and the effect was felt both in sea and on land. The evidence is circumstantial, derived from a comparison of Mesozoic species with closely-related living types; knowing the environmental preferences of existing species, we can infer the climatic conditions 70 million years ago. Nevertheless, the conclusion that the climate was deteriorating seems inescapable.

The worldwide destruction of plankton, ammonites and reptiles in the oceans, and dinosaurs, pterosaurs and plants on land appeared to take place in a catastrophically short time. In nearly all cases their disappearance from the rocks is so abrupt that many authorities are talking in terms of a synchronous decimation of these ubiquitous groups in less than one million years. M. N. Bramlette of the Scripps Institute of Oceanography, in a study of the massive destruction of plankton, estimates that only thousands of years were required for its completion.[14] Others consider that a few days would suffice.

The simultaneous annihilation over the entire earth's surface has led to some, at first sight, bizarre explanations. Cataclysmic theories, ruled out of geology since the age of Baron Cuvier – whose theory of mass extinctions involving inexplicable revolutions or catastrophes coincided rather appropriately with the most vigorous years of the French Revolution – have made a dramatic re-entry. An event that could bring an abrupt end to groups as diverse as dinosaurs and plankton simultaneously across the surface of the globe had to be of truly cosmic proportions.

A supernova explosion, like that observed by Chinese astronomers in 1054, would shower the Galaxy with cosmic radiation of unimaginable magnitude. The explosion of a star of ten solar masses will release a blast equivalent to the combined effects of 1,000,000,000,000,000,000,000,000 ten megaton hydrogen bombs. The medieval Chinese 'guest star' whose remnants we now know as the Crab Nebula was too distant to have led to any radiation hazard for the

Earth. But if a near-by star exploded, one, say, a hundred light years from our solar system, the radiation level at the top of the atmosphere – which is normally about 0·03 roentgens per year – would shoot up to a staggering 3,000 roentgens. Since the radiation would be received in a concentrated form, spread out over a few days at most, the effect would be catastrophic. The radiation level in the atmosphere would increase ten-million-fold and would be similar in effect to a universal nuclear holocaust. Although supernovae are as frequent in our Galaxy as one explosion every fifty years,[15] most, like the Crab, are distant enough to remain harmless. But the Earth does receive an acute radiation dose of 500 roentgens at about 50 million year intervals when a closer star explodes; and probably every 300 million a dose of 1,500 roentgens. Since Precambrian times, 600 million years ago, probability dictates that there should have been one occasion when the Earth was swamped with radiation little less than twice this strength.

The lethal dose for laboratory animals varies from 100 to 700 roentgens, so a radiation blast of supernova magnitude, if it arrived on the planet's surface, would kill many lifeforms. Dinosaurs would have been particularly vulnerable; being large and unprotected, and unable to burrow, they would have been irradiated with massive doses. Those that did not perish outright from cellular disruption and cancer would surely have produced mutant offspring; if, that is, there had been a supernova explosion in the vicinity about 70 million years ago.

There is a great attraction in dabbling with such astronomic figures, especially when they relate to cataclysmic events in the Universe. The morbid appeal of cataclysms lures the mind like spectators to a disaster. But intellectual infatuation is little criterion for acceptance. The anomalies accompanying the theory of direct cosmic radiation are just too great, even supposing that the space scientists have their figures correct. To kill a simple unicellular organism, for example, requires about ten times the dose lethal to a mouse, yet the mammals survived the Cretaceous onslaught whilst plankton were profoundly affected. Since even a thin layer of surface water would effectively blanket out radiation, the theory of direct radiation fails the test because the wrong creatures died. The model has even been criticised by some physicists who insist that the tortuous path taken by the cosmic radiation over ten light years would cause it to arrive stretched out over forty years.[16]

Direct cosmic bombardment may be ruled out. No account, however, had been taken of the temperature fluctuations which were obviously the primary cause of the desolation. At least two workers concerned with dinosaur extinction have been convinced by the arguments of the supernova theorists. Dale Russell of the National Museum of Natural Sciences in Ottawa and Wallace Tucker of American Science and Engineering in Boston have amended the theory to lay equal stress on the climatic effects of impinging radiation. The atmosphere would absorb the equivalent of the power of 100,000 ten megaton hydrogen bombs if a nearby star blew up. The radiation would mostly take the form of X-rays, which would be absorbed by the ozone layer and ionosphere. The ensuing climatic effects would be disastrous. The turbulence generated would drastically alter the heat-retaining property of the atmosphere. The lower, water-saturated air would be displaced into higher regions where ice crystals would form and act to deflect the sun's rays, causing temperatures to fall over the whole globe with alarming suddenness. It is interesting to note in this connection that a sharp increase in X-ray emission from the Sun does result in a shift in atmospheric circulation towards one with definite glacial characteristics.[17] Could supernova bombardment cause the sudden temporary cold spell responsible for wiping out all manner of oceanic and terrestrial life? The dual effects of ground-level radiation resulting from the interaction of cosmic rays and gamma rays with the atmosphere and the lowering of temperatures could account for all the observed changes, according to Russell:

> The events marking the close of the dinosaurian era appear to have been similar to those which may follow the arrival of the initial burst of energy from a nearby supernova. The larger and more highly evolved terrestrial organisms (dinosaurs, flowering plants) would have been decimated by the combined effects of radiation and cold. Marine organisms probably would have been more severely affected by the cooling of the oceans, although some forms inhabiting shallow waters may have been exterminated by radiation or through storm generated water turbulence. The earth was cool sufficiently long to chill the oceans but not long enough to permit the development of continental ice sheets in high latitudes, as there is no evidence of glaciation anywhere on the globe at this time.[18]

This nicely illustrates the severity and abruptness of the devastation envisaged by the supernova theorists. Radiation levels like those following a nuclear war, global storms of extreme intensity, disastrously low temperatures, and the obliteration of the primary elements in the food chain on land (flowering plants and cycads) and in the sea (plankton and ammonites) must have wrought untold havoc. And all this occurring in the space of a week. Why, one wonders, did the mammals survive? And the crocodiles and highly vulnerable birds, all of which were as exposed as the dinosaurs?

There is one obstinate fact, *should it prove to be fact*, that is glaringly inconsistent with the theory of an abrupt synchronous extermination of life. A closer examination of the last dinosaur-bearing rocks seems to show the disappearance as a *progressive* event, albeit a rapid one. (This is disputed by the supernova theorists, who see the apparent staggering as a result of selective prospecting.) The rhinocerine *Triceratops* once roamed the earth in herds like those of the American buffalo before the coming of the Winchester rifle. Its bones are found in the last of the Cretaceous beds, the Lance formation (in, for example, eastern Wyoming[19]). The shales and sandstones of this formation are over 2,000 feet thick from top to bottom, and they leave a record of the creatures that saw out the end of the Mesozoic. *Triceratops*, found in abundance at the bottom of the formation, becomes progressively rarer as we advance upwards to the overlying Tertiary rocks. Even the rare finds eventually cease, marking the exit of the dinosaurs. *Triceratops* must have been the hardiest of dinosaurs. But even its leathery skin was no protection against the approaching winter.

Outliving many of its earlier contemporaries and surviving to the very end of the Cretaceous, *Triceratops* was becoming a lonely figure on the landscape. Colbert in particular has argued that shortly before the close of the Cretaceous a period of decline set in among dinosaurs: not so much a decline in sheer numbers as in variety. The number of dinosaurian genera decreased to such an extent that, as he remarks, the great carnivores must have been suffering from an increasingly monotonous diet.[20] The clichéd confrontation between *Triceratops* and *Tyrannosaurus* was probably less a matter of choice for the giant carnivore than the result of a lack of suitable alternatives. Like almost every other group, Colbert claims, the hadrosaurs began thinning before the end of the period; in fact, the Lance formation has yielded less than a quarter of the number of hadrosaur genera removed from the earlier Belly River formation. The same story can be told of the armoured ankylosaurs, whose Lance representatives in our collections total less than a third of their numbers from earlier beds. By Lance times, ceratopsian genera had apparently been halved in number.[21] If this decline in diversity is real it signifies the early beginning of adverse conditions which acted by weeding out the forms with lowest tolerance. Yet most of the dinosaurs left some hardy survivors which, like *Triceratops*, exploited the lack of competition and expanded into large herds.

In similar fashion, the marine ichythyosaurs seem to have perished before the period's end, and it is possible that plesiosaurs dwindled in the cooling seas before the exit of the last dinosaur on land. These losses at the top of the marine food chain reflect its vanishing plankton base. The ammonites, whose ornate spiral shells had dominated earlier rocks, were also apparently decreasing in diversity before the end, although the rate of extinction drastically increased towards the close of the period.[22] There is a striking parallel between the progressive loss of the marine ammonites and the decrease in dinosaurian variety in the late Cretaceous world. If these findings are upheld they would certainly knock much of the miraculous stuffing out of the arguments of the supernova exponents. Nevertheless, even assuming that these results faithfully reflect the gradual demise of Cretaceous life, one is left with the problem of explaining the extraordinary acceleration of the process of extinction in its final phase. Why should this progressive extinction have been rapidly accelerated? And why did it happen to so many unrelated creatures almost simultaneously, leading to an apparent global crisis? Was there a short-lived but severe cold spell at the very end of the Cretaceous, resulting in a steep gradient being set up between the equator and polar caps, and the falling of the more northerly tropical regions into the temperate zone, as the plankton and conifers testify? Was a steep daily gradient established, with a sharp drop in temperature as dusk approached. Deciduous flowering plants are evidence that the seasons were well-established by late Cretaceous times, so could the seasons have become more marked at the end, with increasingly colder winters?

The terrestrial reptiles that shared the end of the Cretaceous with the dinosaurs would be familiar to us today. By then the crocodiles had a modern aspect. Iguana and monitor lizards flourished, and snakes, though rare, were in existence. Even some of the turtles in the late Cretaceous seas have descendants alive today. They all survived, whereas all the manifold types of dinosaurs – ceratopsians, ankylosaurs, duck-billed hadrosaurs, tyrannosaurs, dromaeosaurs, coelurosaurs, and many more – were truncated simultaneously. What distinguished dinosaurs from reptiles?

The answer is at once apparent: reptiles are cold-blooded whilst dinosaurs were warm-blooded. When the temperature drops below optimum, reptiles become sluggish. They themselves become cold and their metabolic rate drops sharply, driving them into a state of torpor and eventually hibernation. But they remain alive, even at ambient temperatures below freezing, and hence they are able to survive in temperate zones today. It is of no surprise, then, that those reptiles which survived the Cretaceous were also the ones that could crawl under logs to hibernate, or, as in the case of the turtles and crocodiles, tunnel into river banks for protection, somewhat as the Chinese alligator does at the present day. But the dinosaurs were unable to hibernate. Not only were they too large,[23] but they were probably physiologically ill-adapted for it. The reason why is easy to understand. Lizards, with their larger surface area: volume ratio quickly lost heat if the temperature dropped, and this rapidly inactivated them. Lizards had to be prepared to hibernate, but small changes in ambient temperature were easily combatted by the bulky dinosaurs with their proportionately smaller surface area. All the dinosaurs were large, few weighed less than a hundredweight when alive,[24] and some reached 50 tons. Even the young seem to have been born a fairly large size.[25] With such a large body they could easily remain warm in equable conditions and had no need of superficial insulation like hair or feathers. Like large living tropical mammals, such as elephants, hippos and rhinos, dinosaurs needed no protection. The fact that dinosaurs were uninsulated indicates that they had evolved in climatically equable conditions. The mammal-like reptiles, on the other hand, born into the harsh Permian climate, had insulated their bodies with hair and this feature had been inherited by the mammals. It enabled them to shrink in size yet still survive the exposure. It also allowed them to survive the Cretaceous. If world-wide temperatures dropped at the end of the Cretaceous dinosaurs would have been caught out by their own ingenuity. The factor that had spelled success for one hundred and fifty million years would militate against them. They were warm-blooded: they had to maintain a constant internal temperature in order to remain alive, but they lacked insulation against the cold.[26] At first they would have been able to generate enough heat to remain warm, but as the seasons became more marked and frosts became a yearly occurrence even in low latitudes, and the lizards went into hibernation whilst the small fur-covered mammals found themselves warm nests, the dinosaurs would have been left out in the cold. Unable to hibernate because of their unwieldy size, these leviathans would have 'frozen' to death.

The winged pterosaurs pose an altogether different problem. They were furry and warm-blooded like mammals, yet they perished with the dinosaurs. The pterosaur's eventual undoing was the penalty it paid for extreme specialisation. Extensive wings possibly made heat-loss a critical factor. But climatic deterioration would have hit pterosaurs in quite another way. *Pteranodon* was a master of slow soaring, able almost to hang in the air at speeds as low as 15 m.p.h. The equable Cretaceous climate, with uniformly mild weather extending into very high latitudes, would have resulted in light winds ideally suited to soaring. *Pteranodon*, it will be remembered, needed a breeze of only 15 m.p.h. to lift it unaided off the ground. It has been suggested that as the temperature gradient between poles and equator increased, the force of the Westerlies and Trade Winds altered. A rise of only 10 m.p.h. in the prevailing winds would probably have proved too blustery for the gentle glider, causing it to lose control.[27] Lawson's gigantic Big Bend pterosaur would have suffered the additional strain of diminishing food supply. If Lawson guessed correctly and the new pterosaur was a carrion feeder – dependent upon the carcasses of large dinosaurs for its continued existence – its own demise would have followed in the wake of the dinosaur's. It is of great significance that no large-bodied animals survived the Cretaceous; the small inherited the earth. The Texas pterosaur, used to gorging on mountains of rotting flesh, would have been stranded in a world where carrion came rat-size.

Still the supernova theorists remain adamant. Those creatures (the ichthyosaurs specifically) that vanished before the period's end are dissociated from the final catastrophic extinctions. Moreover, Colbert's progressive decrease in diversity among dinosaurs is regarded as illusory. Russell's latest study of dinosaur prevalence in Late Cretaceous times, published in 1975, reveals that indeed there *is* a marked sampling bias: for generations the older Upper Cretaceous beds had been more frequently and more exhaustively prospected than the youngest ones, resulting in a higher yield of dinosaur species. Russell's correlation between the time spent working any formation and the number of species extracted from it is strongly suggestive. Unaware of this sampling anomaly, palaeontologists falsely assumed that dinosaurs were thinning before the end; rather, palaeontologists themselves had collected fewer samples from the latest rocks. Correcting for the bias, Russell determined that *dinosaurian diversity even in the latest Cretaceous was not markedly down.*[28] Not a few stragglers, but the entire complement of dinosaurs had apparently thrived to the end. Could the ammonites have been subject to a similar sampling bias?

According to the latter-day catastrophists, the dinosaurs exited with the most spectacular bang since Creation, and, the geologist's lingering aversion to cataclysms notwithstanding, it is becoming difficult to disagree. The hunt has begun for a nearby supernova remnant and it is encouraging to note that traces of a late Cretaceous stellar explosion have actually been detected.[29] Supernova or otherwise, whatever racked the planet 70 million years ago brought to an abrupt end the Mesozoic world order and closed a major chapter of earth history. Although the timing of the event is still hotly contested, there is little disagreement over its geological suddenness, the debate now focuses on just *how* sudden. The magnitude of the devastation cannot be underestimated, while the obliteration of all large and medium-sized animals coupled with the disruption of marine life profoundly altered the direction that life on earth was to take.

Whatever the nature of the event that we are dealing with, it was of cataclysmic proportions.

Can we ever witness the suffering of the dinosaurs towards the end? What evidence could possibly reveal the anguish of totally alien creatures separated from us by 70 million years? In recent years Heinrich Erben of Bonn University's Institute of Palaeontology has been examining late Cretaceous dinosaur egg-shells, measuring their thickness in different strata and computing trends. Working with thousands of fragments from successive rock layers in Aix-en-Provence and Corbières in the French Pyrenees, Erben's statistical analysis has had stunning and unforeseen results. The rock layers containing the fossil shells represent the upper parts of the Upper Maestrichtian, and thus cover the short period of time immediately prior to the disappearance of the dinosaurs. Eggs from the older layers are thick-shelled – anything up to 2·5 mm, perfectly acceptable in eggs of large size. But on moving to higher (and thus younger) strata Erben found the shells becoming progressively thinner, whilst those from the youngest rocks are so pitifully thin – often only 1 mm – that they must have suffered from extreme fragility.[30]

Since Erben found many whole eggs, we can assume that they survived the ordeal of being laid; but these same unbroken shells also indicate that the eggs had failed to hatch. The implications are obvious. Dinosaurs were reacting to a brief but protracted[31] period of stress in the same way as birds today, by laying eggs with ever-thinning shells. Stress in birds, whether from cold, DDT poisoning or overcrowding, results in an inbalance of the delicate hormonal system which affects clutch size and shell thickness. The shells may be so thin that they crack, or lack sufficient calcium for the embryos to absorb and use to build their skeletons. In the last few years of its dominion the dinosaur was subjected to unbearable pressure. That much is certain. Those solitary beasts that managed to struggle on left us a glimpse of the agony that they must have suffered. With their hormonal systems hopelessly out of tune, the dinosaurs reacted by laying weak-shelled eggs, and in so doing sealed the fate of their own offspring.

In 1973 the Bonn team confirmed their earlier findings when they located eight more eggs – two of them intact – in a rock face near Corbières. The unbroken eggs were large, about nine inches long and seven inches broad, and came from the youngest horizon. The fragments, viewed through an electron microscope, were so poorly shelled that the dinosaur embryos could not possibly have absorbed enough calcium to complete the building of their skeletons. The majestic dinosaurs, Mesozoic overlords that anticipated the mammalian or avian level of organisation, had departed not with a bang, but with a whimper – the whimper of the young as they perished incarcerated in tiny prisons.

Geological Time and Vertebrate History

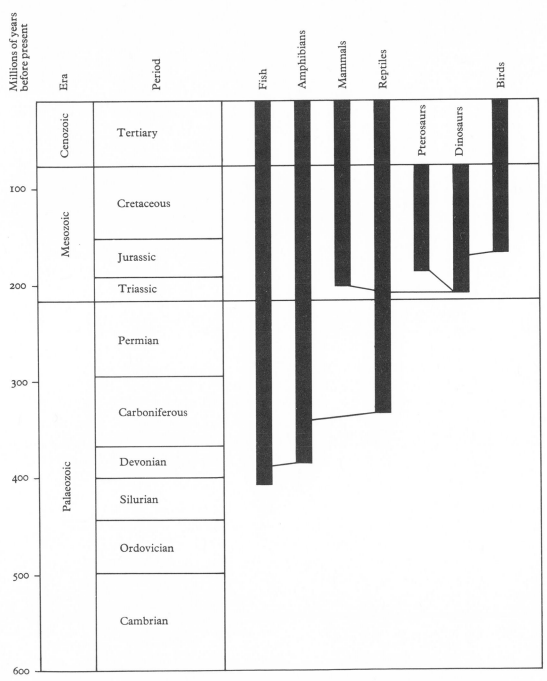

A glossary of technical terms appears on page 204 and a family tree of the higher vertebrates concludes Appendix 1.

Appendix 1 : A Note on Classification

The Mesozoic animals that play a crucial role in the present thesis – dinosaurs, thecodonts and pterosaurs – are still officially termed reptiles. The reptilian class was originally erected to house the living snakes, lizards and turtles long before the re-emergence of the Mesozoic saurians in the early 19th century. As such, the reptiles were, by definition, cold-blooded with a scaly covering and sprawling limbs, unable to sustain energetic activity, and generally confined to climatically equable regions. Cuvier justified the inclusion of pterodactyls, and Owen the dinosaurs, on the basis of some apparently lizard-like bones in the fossil skeletons. Given the state of contemporary knowledge, this attempt to ally past and present lifeforms was laudible for its heuristic value.

But palaeontology has not stood still for the intervening century or more. Slowly, sometimes almost imperceptibly, more often painfully, old ideas have been eroded piecemeal: witness the growing awareness of the pterosaur's avian-style endothermic physiology, its large brain and hairy pelt; the controversies over the dinosaur's mammalian stance and the subsequent correlation of this with warm-bloodedness; and the thecodont's erect posture and active metabolism. Yet the classification has failed to keep pace with our changing ideas, resulting in an anomalous situation whereby the reptilian class contains creatures as diverse as furry, intelligent, endothermic flyers; archaic, sprawling, small-brained ectotherms (lizards and turtles); giant, erect, terrestrial endotherms; and swift-running, extremely sophisticated dinosaurs, at least one (*Archaeopteryx*) with feathers. The arbitrary nature of this grouping is proving an embarrassment to many palaeontologists.

Classification should not only mirror the evolutionary progress of life, but also reflect the acquisition of major new levels of organisation and their subsequent exploitation. Current 'higher' vertebrate classification fails miserably to reflect the achievements of the dinosaurs and pterosaurs.

The Reptilia is a mixed bag of singularly diverse forms, and although not originally intended, it has become an arbitrary collection of animals. Clearly, something must be done.

One of the major discontinuities in the history of life occurred in Early Triassic times – about 210 million years ago – as the small pseudosuchian thecodonts first achieved an erect stance, lengthened their hind limbs to increase the stride, and stepped up their energy producing ability. This suite of characters was a key innovation of major importance and consequence, for it allowed their descendants to conquer the land and air. It triggered the dinosaurian and

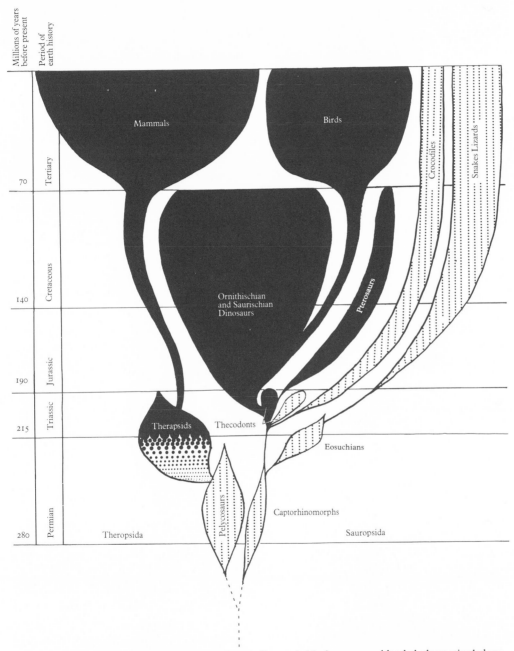

A diagrammatic family tree of the higher vertebrates. Groups in black are warm-blooded, those stippled are cold-blooded. Endothermy evolved twice during vertebrate history, among therapsids and thecodonts. Dinosaurs and pterosaurs inherited the condition from thecodonts, and it passed to birds via the dinosaurs.

pterosaurian explosions.

This must be recognised in the classification. Allying dinosaurs and lizards can no longer be justified. The warm-blooded Mesozoic 'reptiles' had broken with their ancestors in the same way as the mammals had broken with theirs. Because mammals survive, we are quickly made aware of their distinction and accord them a class of their own. The same is true of the birds. Dinosaurs and pterosaurs – despite the expanse of their reign – failed to survive, and had originally to be classified from fragmentary remains and inadequate knowledge. Only now are we becoming aware of their uniqueness.

The warm-blooded 'saurians' of the Mesozoic world should be accorded a rank to match their physiological, anatomical and ecological importance. Therefore, the Classes Dinosauria (including the Pseudosuchia and Aves) and Pterosauria are proposed. Since all these animals inherited their physiological make-up from the early Triassic thecodonts, dinosaurs and pterosaurs may be grouped together naturally in the Superclass Endosauropsida.

Appendix 2 : Some Problems

'Criticism of our conjectures is of decisive importance : by bringing out our mistakes it makes us understand the difficulties of the problem which we are trying to solve.'

Karl Popper, *Conjectures and Refutations*.[1]

The foregoing is not intended in any way as a definitive statement. That is the unfortunate lot of the text-book. It is merely an interim report of the state of the science – and palaeontology is in great flux. Perhaps a few of the ideas and speculations will be corroborated, but inevitably many will prove incorrect. Already certain problems have come to light. These, however, detract little from the explanatory power of our endothermic dinosaur theory. Nevertheless, these difficulties require an airing, and it is hoped that this will prove a stimulus to further research. In short, then, and in accordance with the tenets of Popperian philosophy, we must expose our theory to the severest criticism in order to gauge its worth.

One of the strongest criticisms levelled against the theory concerns dinosaurian brain size, and it was first formulated by Alan Feduccia in 1972. Birds and mammals possess relatively large brains, in marked contrast to the notoriously small brains of most dinosaurs. Yet Ostrom and Bakker postulate advanced levels of activity like those found in mammals and birds for dinosaurs. Surely, Feduccia argues, such levels of activity would have resulted in an increased complexity of the dinosaurian brain.[2] Unknown to Feduccia at this stage, Dale Russell had demonstrated quite conclusively that some dinosaurs, in fact the most sophisticated, the dromaeosaurs and struthiomimids, *did* evolve relatively gigantic brains (some even larger than those of birds) to coordinate balance, agility, visual acuity and manual dexterity, so these must be exempted from the objection.[3] Indeed, Feduccia later conceded 'perhaps they [meaning Russell's ostrich dinosaurs and dromaeosaurs] were endotherms, or at least smart for the Cretaceous'.[4] But this still leaves the problem of the disparity between brain and body size in the majority of large dinosaurs. Could it be that, although the small, gracile dinosaurs required large brains, the stately giants lacked the sophisticated behaviour that demanded an enlarged brain, yet at the same time they were warm-blooded? Endothermy *per se* may not require a large brain. Harry Jerison, in his exhaustive study of the *Evolution of the Brain*, has only recently suggested that 'we might expect, *a priori*, that animals with extrinsic temperature control, such as lizards, which must make behavioural responses moving into or out of sunny or shady areas . . . would use *more* neural tissue than [endotherms] in handling the control of body temperature'.[5] In other words, the sun-orientated behavioural responses of lizards may require more brain volume than the purely automated internal temperature regulating mechanism of an endotherm. On this supposition a warm-blooded brontosaur would need a smaller volume of brain than a heliotropic reptile to deal with temperature control. Jerison warns against jumping to the conclusion that endothermy itself demanded an enlarged brain. Controlling a warm-blooded physiology requires only a very small area of the hypothalamus – dinosaurs could be endotherms even with diminutive brains. Similarly, therapsids probably evolved a warm-blooded physiology in later Permian times, even though they retained small brains which failed to fill their braincases. Early (Triassic) mammals inherited these small brains, and only much later did the brains expand. Again, endothermy seems to have evolved before the enlarged brain.

The evolution of the enlarged brain is better correlated with the development of behavioural sophistication. It is by no means coincidental that the large-brained pterosaurs and dromaeosaurs were remarkably sophisticated in their respective ways. Rather than endothermy demanding a large brain, the converse probably holds: only a fast-metabolising endotherm can realise the complex behaviour we find in, for example, the intricate flight manoeuvres of pterosaurs or the manipulative and balancing ability of dromaeosaurs, and this complex behaviour evolves concomitantly with its coordinating controls in the brain. Endothermy may be one of the prerequisites for brain expansion, not vice versa.

A related problem, though not crucial to the general thesis, lies in the attempted reconciliation of the herding tendencies of dinosaurs with their small brains. Struthiomimids and dromaeosaurids were undoubtedly capable of social organisation (perhaps including an avian-style peck order), parental care, and so on, but does *Triceratops'* minute brain militate against it being a herding animal?

A *leit-motif* of the book has been the inextricable relationship existing between endothermy and dinosaurian size: uninsulated dinosaurs were forced to be titans, and they were thus excluded from the microniches that fell to mammals and lizards. But if the dinosaur was unable to shrink below a critical size, how are we to explain its offspring beginning life below this threshold level without perishing from exposure? One dinosaur that is unquestionably associated with its eggs is the small Mongolian *Protoceratops*. It is significant that the eggs are very large (far larger relatively than those of reptiles), so the offspring began life larger than many contemporary mammals and lizards. The timing of hatching would also have coincided with the mildest season to reduce the severity of the exposure.[6] Whether these arguments are strong enough to meet the objection remains to be seen.

Warm-bloodedness is expensive to maintain; it requires the burning of a great deal of fuel. A man consumes about forty times as much food as a man-sized lizard, and about 80–90% of the energy released is utilised in maintaining a constant temperature. Therefore, a brontosaur would require an order of magnitude more food than a brontosaur-sized lizard: perhaps as much as half a ton of vegetation per day. This raises the intriguing question of just how so much food could have been passed through a head that was, in the case of *Brontosaurus*, only vertebra-sized. Then, to release the energy, the food had to be mechanically broken down, which renders all the more perplexing the lack of grinding teeth; *Diplodocus* possessed only a few nipping 'incisors'. Presumably, gizzard stones substituted for crushing teeth.

Finally, although again not crucial to the central thesis, comes the nagging problem of the lack of *any* dinosaurs surviving into post-Cretaceous times. If cold was responsible for their annihilation, even if it accompanied the initial burst of radiation from a nearby supernova, why did dinosaurs not persist in the warmer equatorial regions? Is it conceivable that the *entire* globe was subjected to intense cold? It seems unlikely. Even though both land masses in the northern hemisphere, America and Eurasia, were situated in middle and high latitudes in late Cretaceous times, and were bordered to the south by the Tethys Ocean, South America and Africa were more favourably situated climatically. The abruptness, ubiquity and severity of the Cretaceous extinctions makes it increasingly difficult to dismiss a cataclysmic theory.

Glossary

aetosaur	Triassic thecodont reptile with a pig-like snout. Terrestrial four-footed root-grubber.
air sacs (lung sacs)	An extension of the lungs throughout the body cavity of birds, pterosaurs and some dinosaurs. These air-filled sacs may enter the bones.
alkaloid	Organic chemical synthesised by flowering plants (includes morphine and strychnine).
ammonite	Tentacled mollusc related to the squid, but with a coiled and chambered shell. Plentiful in Mesozoic seas.
ankylosaur	Tank-like armoured dinosaur. A quadrupedal ornithischian.
Antediluvian	A term commonly used in the eighteenth and early nineteenth century to describe the artefacts (especially fossils) that were supposed to have predated Noah's Flood.
archosaur	A grouping of higher vertebrates based on certain shared skeletal features. Comprises dinosaurs, pterosaurs, birds, crocodiles and thecodonts.
Aves	Birds.
batrachian	Amphibian.
belemnite	Tentacled mollusc related to squid, but with bolt-shaped guard or shell.
biped	An animal that stands and walks on its hind legs.
brontosaur	Giant quadrupedal saurischian dinosaur (e.g. *Brontosaurus*).
camptosaur	Primitive ornithopod dinosaur from the Upper Jurassic of North America.
ceratopsian	Horned ornithischian dinosaur. The group flourished in the late Cretaceous (e.g. *Triceratops*, *Protoceratops*).
cerebellum	Region of the vertebrate hindbrain, concerned with co-ordination of movement.
cerebral hemispheres	Paired outgrowths in the vertebrate forebrain, a higher coordinating centre.
cervical vertebra	Neck vertebra.
cetacean	Whale.
cetiosaur	Sauropod.
clavicle	Collar bone.

coccolith	Microscopic, single-celled planktonic plant with a hard shell. On the death of these minute organisms, the shells drop to the sea bed and eventually compact to form chalk.
coelurosaur	One of the theropod dinosaurs. Coelurosaurs were bipedal, fleet-footed dinosaurs with a delicate build. Ornithomimids are known to have had larger than average brains. These dinosaurs were bird-like in appearance and actually ancestral to birds. (e.g. *Ornithomimus*, *Compsognathus*.)
cold-blooded	Animals whose internal temperature may fluctuate with the ambient temperature. With a cold external temperature the internal temperature drops and the animal becomes sluggish, although by basking in the sun cold-blooded creatures can raise their temperature above that of the air. (e.g. lizards, crocodiles, amphibians.) See ectothermy.
convergence	Creatures that have come to look similar, even though their ancestors may have been totally different (and only distantly related), a good example being the mammalian dolphin and reptilian ichthyosaur. Animals evolving similar structures do so in response to similar demands of lifestyle.
Cretaceous	Last period of Mesozoic, 140–70 million years ago.
cycad	Palm-like plant possessing a single stem with a crown of fern-like leaves, flourishing in the Triassic and Jurassic. Represented by very few surviving genera.
cursorial	With slender limbs adapted to running, in contrast to graviportal.
Danian	Earliest Tertiary beds.
Devonian	A period of the Palaeozoic, 395–345 million years ago. Emergence of the first fish on to land.
dicynodont	Therapsid or mammal-like reptile. Dicynodonts were corpulent herbivores abundant in the Permian and Triassic, possibly living in herds. Many possessed walrus-like tusks.
dromaeosaurid	One of the theropod dinosaurs. Little known. Very agile, with a large brain and huge eyes. Mid to Late Cretaceous. (e.g. *Deinonychus* and *Saurornithoides*.)
ectothermy	Reliance on an external heat source (the sun) for raising the internal temperature. See cold-blooded.
endothermy	Reliance on internal heat sources; heat generated by the muscles and tissues. The animal remains warm despite fluctuations in ambient temperature. See warm-blooded.
Eocene	Epoch in the Tertiary period, 55–37 million years ago.
femur	Thigh bone.
foraminiferan	Single-celled planktonic organism with a hard shell.

fossil	Petrified remains of an animal or plant. As rock-forming sediments accumulate, they trap the hard parts of animals and plants (and sometimes, if the sediments are fine enough, traces of softer tissues), thus preserving them.
gastrolith	Stone used to grind food in the gizzard or stomach.
genus (pl.genera)	Unit of classification. A collection of related species. *Diplodocus* is the genus, *Diplodocus carnegiei* the species. There may be more than one *Diplodocus* species.
gizzard	Pre-stomach in birds and possibly dinosaurs. Has a strong muscular wall; gastroliths sometimes present. Food may be swallowed unchewed (as it is in birds) and the gizzard grinds and churns the food to a pulp before passing it to the stomach.
graviportal	Heavy animals constructed sturdily, often with pillar-like limbs. The gait is usually an amble. (e.g. elephants and brontosaurs.)
humerus	Upper arm bone.
ichthyosaur	Marine, dolphin-shaped reptile living in Mesozoic times.
iguana	Tropical lizard.
iguanodont	Bipedal, herbivorous, ornithischian dinosaur.
Jurassic	The middle period of the Mesozoic, 190–140 million years ago.
labyrinthodont	Squat, four-footed armoured amphibian, common in the Permian and Triassic.
Lamarckian	Mechanism of evolution promulgated by Lamarck (1744–1829). He believed that increased use of a structure caused it to grow. Now replaced by neo-Darwinism, involving mutations and natural selection.
Lias	Early Jurassic.
Maestrichtian	Latest Cretaceous beds.
marl	Muddy limestone.
marsupial	Pouched mammal (e.g. opossum, kangaroo, marsupial wolf.)
medulla	Most posterior part of the vertebrate brain.
Mesozoic	One of the main geological eras, 215–70 million years ago, and made up of three periods, the Triassic, Jurassic and Cretaceous. Referred to colloquially as the 'Age of Reptiles'.
metabolism	This refers to the chemical processes within the body, especially the breakdown of food to release energy. A fast metabolism entails the ability to release more energy and with greater speed. This involves an increased uptake of oxygen to 'burn' the food, as well as a quickening of the chemical reactions in the tissues. The rate of metabolism is dependent upon temperature; reactions are slower if the temperature drops below the optimum for the catalysing enzymes. Most modern reptiles are capable of

short bursts of fast metabolism, but they are unable to sustain it. Mammals and birds, in contrast, can sustain this increased energy output and thus are capable of a higher level of activity.

metacarpal — Long bone in the palm of the hand, between the wrist and digit.

metatarsal — Long bone in the sole of the foot, between the ankle and digit.

moa — Large, flightless bird, now extinct, Australasian (e.g. *Dinornis*).

monitor — Tropical or subtropical lizard living in Old World, may grow to a large size (e.g. Komodo dragon in the East Indies). Related to the Cretaceous mosasaur.

monophyly — Single origin for a group of animals (birds are probably monophyletic, mammals are not: monotremes evolved from therapsids independently of placentals and marsupials).

monotreme — Very primitive mammal, egg-laying, and with little ability to lose heat. Evolved independently of other mammals from therapsids. e.g. duck-billed platypus.

New Red Sandstone — Permian.

niche — Type of environment preferred by an animal. e.g. open plains for long-legged runners, undergrowth for shrews, and so on.

olfactory bulb — Region of the vertebrate forebrain concerned with smell.

Oligocene — Epoch of the Tertiary period, 37–27 million years ago.

Oolite — Middle Jurassic. Oolite means 'egg stone', and many of the pale-coloured limestones in the series are composed of small grains like fish eggs. Jurassic Oolite rocks extend across England, from Yorkshire in the north-east to Dorset in the south-west. The Stonesfield Slate, housing the *Megalosaurus*, is part of the Oolite series.

opisthotonic — Nervous disruption by poison, causing muscular spasms and often death. Muscles tighten and the animal dies in a contorted pose. Can be caused by alkaloids such as strychnine.

optic lobe — Region of the vertebrate midbrain concerned with vision.

Ornithischia — One of the two major groups of dinosaurs; comprising ornithopods, stegosaurs, ankylosaurs and ceratopsians.

ornithopod — Bipedal herbivorous ornithischian dinosaur (e.g. *Iguanodon*, *Hadrosaurus* and *Hypsilophodon*).

pachyderm — Large quadrupedal animal (e.g. elephant and rhino).

palaeontology — The study of past life as revealed by the fossil record, involving the reconstruction of the appearance of ancient organisms, their mode of functioning, ecology and affinities to other creatures.

Permian — Last Palaeozoic period, 280–215 million years ago. Gave way to the Triassic.

pes	Foot.
phalanx (pl.phalanges)	Finger or toe bone.
physiology	The science dealing with the processes of living matter and the functions of organs and tissues.
phytosaur	Aquatic Triassic thecodont with a long crocodile-like snout.
placental	Mammal with a placenta (all surviving mammals except marsupials and monotremes).
plankton	Small marine organisms of diverse type that drift with the currents.
plantigrade	Walking on the soles of the feet; in contrast to walking on the toes (digitigrade).
plesiosaur	Marine or freshwater Mesozoic reptile with paddles.
Pliocene	Last epoch of the Tertiary period, 12–2 million years ago.
pneumatic bone	Hollowed bone, filled in life by lung sac. Found in pterosaurs, birds and some dinosaurs.
polyphyly	Multiple origins for a group of animals: cf. monophyly.
pre-Adamite	Term common in eighteenth and early nineteenth century, applied to animals and plants supposedly existing before Adam.
preadaptation	The development of a structure in one environment (e.g. lungs in a fish) that proves to be of even greater use in another (e.g. on land).
prosauropod	Primitive Triassic saurischian dinosaur. Includes both bipedal and quadrupedal forms.
pseudosuchian	A Triassic thecodont. Very often bipedal, although there were quadrupedal forms. Here we find the development of a fully erect posture for the first time in vertebrate history. Ancestral to dinosaurs.
pteroid bone	Small bone arising from the wrist in pterosaurs. It probably supported a membrane to the neck that acted as an elevator.
quadruped	Animal that stands and walks on all fours.
radius	Forearm bone.
Roentgen	The roentgen, named after the discoverer of X-rays Wilhelm Röntgen (1845–1923), is the standard measure of radiation. Life on earth is exposed annually to a little over 0.1 roentgens, which comes from the soil, atmosphere and even breakdown of radioactive potassium and carbon in living tissues. Cosmic radiation takes the form of high energy X-rays and gamma rays; in excess these cause cellular breakdown in tissues and this can lead to cancer. Reproductive cells are particularly vulnerable and irradiation can cause sterility or the birth of mutants.

sacrum	Vertebrae in the pelvic region, fused to one another and attached to the pelvic girdle.
saurian	Reptile (early nineteenth century usage).
Saurischia	One of the two main groups of dinosaurs, comprising theropods (all flesh-eaters) and sauropods (brontosaurs, etc.).
sauropod	Colossal quadrupedal saurischian dinosaurs, flourishing in the Jurassic but surviving to the end of the Cretaceous (e.g. *Brontosaurus* and *Diplodocus*).
secondary palate	Bone forming the roof of the mouth in mammals, crocodiles and some dinosaurs, separating the mouth from the nasal cavity.
species	The smallest unit of classification in common use. *Diplodocus carnegiei* is a species. There may be many species in a genus (*Diplodocus* is the genus).
sprawling	Limbs held out at the sides of the body, with the upper leg bones (femur and humerus) parallel to the ground.
stegosaur	Quadrupedal ornithischian dinosaur with a series of bony plates along the ridge of the back.
sternum	Breastbone.
Stonesfield Slate	Middle Jurassic.
supernova	The explosion of a star.
tannin	An organic chemical synthesised by plants. Common in the bark of trees.
Tertiary	Period following the Cretaceous. Referred to coloquially as the 'Age of Mammals'. 70–2 million years ago.
Tethys ocean	An ocean extending in the east-west direction during the Mesozoic. Extensive in area and separating the northern and southern supercontinents.
thecodont	Triassic reptile with socketed teeth; pseudosuchians, aetosaurs and phytosaurs are included.
therapsid	Mammal-like reptile. Among therapsids were the ancestors of mammals. Therapsids were the dominant land animals in later Permian and lower Triassic times, but were ousted by the pseudosuchian thecodonts.
theropod	Flesh-eating saurischian dinosaur. The group comprises carnosaurs (e.g. *Tyrannosaurus*), coelurosaurs (e.g. *Ornithomimus*) and dromaeosaurs (e.g. *Deinonychus*).
tibia	Shin bone.
Triassic	First period of Mesozoic, 215–190 million years ago.
ventricle	Chamber in the heart responsible for pumping blood to the body and lungs.
warm-blooded	An animal that maintains a high internal temperature, regardless of external fluctuations, by generating its own heat. Mammals, birds, dinosaurs, pterosaurs, and some thecodonts. See endothermy.

Wealden

Lower Cretaceous. The Wealden derives its name from the 'Weald' of Kent, where it was first studied. It is composed of sandstones and clays, originally deposited in Early Cretaceous times, over 100 million years ago, in a huge fresh or brackish water estuary extending across southern England. *Iguanodon* and *Hylaeosaurus* lived in Wealden times.

General Bibliography

The works listed differ greatly in form and content, covering a spectrum of subjects ranging from descriptive anatomy to specific historical questions.

Edwin H. Colbert, 1962. *Dinosaurs : Their Discovery and their World.* Hutchinson, London.

Edwin H. Colbert, 1965. *The Age of Reptiles.* Weidenfeld & Nicolson, London.

Edwin H. Colbert, 1969. *Evolution of the Vertebrates.* 2nd ed., Wiley, New York.

Edwin H. Colbert, 1971. *Men and Dinosaurs : The Search in Field and Laboratory.* Penguin.

Edwin H. Colbert, 1974. *Wandering Lands and Animals.* Hutchinson, London.

Loren Eiseley, 1961. *Darwin's Century : Evolution and the Men who Discovered it.* Anchor Books, New York.

Charles Coulston Gillispie, 1959. *Genesis and Geology : A Study in the Relations of Scientific Thought, Natural Theology, and Social Opinion in Great Britain, 1790–1850.* Harper Torchbooks, New York.

John C. Greene, 1959. *The Death of Adam : Evolution and its Impact on Western Thought.* Ames, Iowa State UP.

L. B. Halstead, 1969. *The Pattern of Vertebrate Evolution.* Oliver and Boyd, Edinburgh.

Gerhard Heilmann, 1926. *The Origin of Birds.* H. F. and G. Witherby, London.

Url Lanham, 1973. *The Bone Hunters.* Columbia University Press, New York and London.

W. D. Matthew, 1915. *Dinosaurs.* New York, American Museum of Natural History Handbook.

Alfred Sherwood Romer, 1966. *Vertebrate Palaeontology.* 3rd ed., University of Chicago Press, Chicago and London.

Alfred Sherwood Romer, 1971. *Man and the Vertebrates.* 2 Vols., Penguin.

M. J. S. Rudwick, 1972. *The Meaning of Fossils : Episodes in the History of Palaeontology.* Macdonald, London; American Elsevier, New York.

Harry Govier Seeley, 1901. *Dragons of the Air.* Methuen, London. (Reprinted 1967, Dover paperback.)

Barbara J. Stahl, 1974. *Vertebrate History : Problems in Evolution.* McGraw-Hill, New York.

W. E. Swinton, 1934. *The Dinosaurs : A Short History of a Great Group of Extinct Reptiles.* Thomas Murby & Co., London.

W. E. Swinton, 1965. *Fossil Amphibians and Reptiles*. 4th ed., British Museum (Natural History).

W. E. Swinton, 1965. *Fossil Birds*. 2nd ed., British Museum (Natural History).

W. E. Swinton, 1969. *Dinosaurs*. 4th ed., British Museum (Natural History).

W. E. Swinton, 1970. *The Dinosaurs*. George Allen & Unwin Ltd., London; John Wiley & Sons, Inc., New York.

Herbert Wendt, 1968. *Before the Deluge*. Doubleday & Co., New York.

Notes and References

1. The crown of creation

1 Faujas-Saint-Fond, B., 1799. *Histoire Naturelle de la Montagne de Saint-Pierre de Maestricht*. Paris, pp. 59–67. This account of the Maestricht mosasaur was written during the revolutionary era by a professor of geology at the Paris museum so it naturally champions the scientist Hoffmann and casts the cleric Godin as the villian.

2 For an account of Cuvier, Lamarck and extinction see: Rudwick, M. J. S., 1972. *The Meaning of Fossils*. London and New York. Cuvier's zoological work is dealt with by Coleman, W., 1964. *Georges Cuvier : Zoologist*. Cambridge, Mass. The concept of the plenum and its influence on scientific thought, especially as it touches on the notion of extinction, may be found in: Lovejoy, A. O., 1964. *The Great Chain of Being*. Cambridge, Mass., chapters 8 and 9.

Cuvier demonstrated not only the mastodon's disappearance but went on to investigate the Paraguayan *Megatherium* or giant ground sloth and the Siberian mammoth. There had, suggested Cuvier, been a violent revolution in the earth's recent past that had wiped out much of the animal life. When Cuvier became aware of the more remote Mesozoic reptiles he modified his theory to one of repeated catastrophes, and following each one a new fauna of more modern aspect was introduced. Cuvier's progressionism had to be amended so many times with the discovery of increasing numbers of extinct animals that by 1849 29 catastrophes were needed. Of course, each fresh creation came to look more and more like the next and the whole edifice finally broke down under its own weight. It was superseded by Darwin's theory of evolution in 1859. See: Bourdier, F., 1969. Geoffroy Saint-Hilaire versus Cuvier: the campaign for paleontological evolution (1825–1838). In Schneer, C. J. (ed.), 1969. *Toward a History of Geology*. Cambridge, Mass., pp. 36–61. Esp. p. 60.

3 Cuvier, G., 1836. *Recherches sur les Ossimens Fossiles*. 4th ed., Paris, p. 175.

4 Buckland, W., 1837. *Geology and Mineralogy*. 2nd ed., vol. 1, p. 217.

5 Ibid. p. 220.

6 Buckland, W., 1824. Notice on the Megalosaurus or great fossil lizard of Stonesfield. *Trans. Geol. Soc*. London, series 2, vol. 1, pp. 390–396.

7 Mantell, G., 1825. Notice on the Iguanodon, a newly discovered fossil reptile, from the sandstone of Tilgate forest, in Sussex. *Phil. Trans. Roy. Soc*., 115, pp. 179–186. Mantell's views on these great 'Fossil Lizards', as he called them, is to be found in: Mantell, G., 1833. *The Geology of the South East of England*. London. This book also contains the first published account of the *Hylaeosaurus* unearthed the previous year in Tilgate Forest.

8 Buckland, W., 1835. On the discovery of fossil bones of the Iguanodon, in the iron sand of the Wealden formation in the Isle of Wight, and in the Isle of Purbeck. *Trans. Geol. Soc*. London, series 2, vol. 3, pp. 425–432.

9 Eleven years earlier, in 1830, Hermann von Meyer had gone some way to anticipate Owen. Meyer had split up the fossil reptiles and grouped *Megalosaurus* and *Iguanodon* together because of their pachydermal feet. He defined these 'Pachypoda' as 'Saurians with locomotive extremities like those of the bulky terrestrial Mammals'. Owen, R., 1874–1889. *Monograph of the Fossil Reptilia of the Mesozoic Formations*. Palaeontographical Society, pp. 201, 595 *et seq*. Owen's great protagonist Thomas Henry Huxley insisted on giving Meyer the credit for appreciating the importance of the group, much to Owen's annoyance. In fact, as Owen argued, Meyer's was merely a tabular arrangement and Meyer had not published his reasons for grouping the dinosaurs together. Owen, on the other hand, had gone to great lengths to elucidate the anatomical differences between dinosaurs and other reptiles.

10 Owen, R., 1841. Report on British Fossil Reptiles. *Report of the Eleventh Meeting of the British Association for the Advancement of Science*, pp. 60–204. See especially p. 103.

11 Ibid. pp. 142–143.

12 Ibid. p. 110.

13 Ibid. p. 200.

14 Lamarck, J.-B., 1809. *Philosophie Zoologique*. Paris. Translated into English as *Philosophical Zoology* by Elliot, H., (1914, London; 1963, New York). See Preface p. 2 of Elliot's translation.

15 Owen, op. cit. (10), p. 196.

16 Ibid. p. 202.

17 Ibid. p. 204 (footnote).

18 Desmond, A. J., 1974. Central Park's fragile dinosaurs. *Natural History*, 83 (8), pp. 64–71. For primary sources see Hawkins, B. W., 1854. On visual education as applied to geology. *Journal of the Society of Arts*, 2, pp. 444–449. Owen, R., 1854. *Geology and Inhabitants of the Ancient World*. London.

19 *The Illustrated London News*, Dec. 31, 1853, No. 661, Vol. 23, pp. 599–600. [The Crystal Palace at Sydenham.]

20 Owen, op. cit. (18).

21 *The Illustrated London News*, Jan. 7, 1854, No. 662, Vol. 24, p. 22. [The Crystal Palace at Sydenham.]

22 *The London Quarterly Review*, 1854, 3 (5), pp. 232–279. 'The Fossil Dinner' and appraisal of the monsters is found on pp. 237–239. For the popular reaction see *The Westminster Review*, 1854, 62 (6), pp. 540–541.

23 Dickens, C. J. H., 1853. *Bleak House*, London.

24 Bakker, R. T., 1968. The superiority of dinosaurs. *Discovery* (New Haven), 3 (2), pp. 11–22. See p. 11.

25 Bogert, C. M., 1959. How reptiles regulate their body temperature. *Scientific American*, 200 (4), pp. 105–120.

26 Ibid.

27 Colbert, E. H., Cowles, R. B., and Bogert, C. M., 1946. Temperature tolerances in the American alligator and their bearing on the habits, evolution and extinction of the dinosaurs. *Bull. Am. Mus. Nat.Hist.*, 86, pp. 331–373.

28 Colbert, E. H., Cowles, R. B., and Bogert, C. M., 1947. Rates of temperature increase in the dinosaurs. *Copeia*, 2, pp. 141–142.

29 Colbert, E. H., 1962. *Dinosaurs : Their Discovery and their World*. London, p. 207.

30 Bogert, op. cit. (25).

31 Ibid.

2. The tyrant finds its feet

1 Hayden, F. V., 1860. Geological sketch of the estuary and fresh water deposit of the bad lands of the Judith, with some remarks upon the surrounding formations. *Trans. Am. Philos. Soc.*, 11, pp. 122–138.

2 Leidy, J., 1860. Extinct vertebrata from the Judith River and great lignite formations of Nebraska. Ibid. pp. 139–154. This paper, read before the American Philosophical Society in 1859, appeared after Leidy's description of *Hadrosaurus*, even though Leidy had acquired the *Deinodon* and *Trachodon* some months earlier than the *Hadrosaurus* skeleton; their significance was obviously increased by the discovery of the skeleton.

3 Leidy, J., 1858. *Hadrosaurus* and its discovery. *Proc. Acad. Nat. Sci.*, Philadelphia, Dec. 14, pages 213–218. See p. 217.

4 Ibid. p. 217.

5 Letter from Cope to his father, dated August 15, 1866. Reproduced in Osborn, H. F., 1931. *Cope : Master Naturalist, The Life and Letters of Edward Drinker Cope*, Princeton, p. 157.

6 Cope, E. D., 1866. Remains of a gigantic extinct dinosaur. *Proc. Acad. Nat. Sci.*, Philadelphia, Aug. 21, pp. 275–279.

7 Board of Commissioners of the Central Park, 1868. *Twelfth Annual Report*, Appendix F, pp. 125–141.

8 Waterhouse Hawkins' father, it seems, had been a close friend of Charles Willson Peale, the founder in 1786 of the Philadelphia museum of natural history. Early in the nineteenth century Peale discovered and rebuilt the first complete skeleton of an American mastodon, and in 1802 many of Peale's friends joined him *in a dinner inside the skeleton of the 'mammoth'*. Among those present was Hawkins' father, an inventor whose 'Patent Portable Grand Piano' was brought along for the festivities. That year Hawkins' father returned to his native England. The uncanny resemblance between this episode in Philadelphia and Waterhouse Hawkins' own dinner inside the *Iguanodon* at Crystal Palace 52 years later leads us to the conclusion that Waterhouse Hawkins heard of the affair from his father. For this information I am indebted to Peale's biographer, Charles Coleman Sellers. Pers. comm. Nov. 3, 1974.

9 Desmond, A. J., 1974. Central Park's fragile dinosaurs. *Natural History*, 83 (8), pp. 64–71.

10 Marsh, O. C., 1890. Reply to Professor Cope. *New York Herald*, Jan. 19.

11 Marsh, O. C., 1877. Notice of a new and gigantic dinosaur. *Am. Jour. Sci.*, 14, pp. 87–88.

12 Cope, E. D., 1877. On the Vertebrata of the Dakota Epoch of Colorado. *Proc. Amer. Philos.Soc.*, 16, pp. 233–247.

13 Olmstead has subsequently become the patron saint of landscape gardening. He was responsible for most of the great projects of his time, ranging from Harvard Yard to Central Park. Like many others with enlightened views he fell foul of the Tweed administration.

14   Callow, A. B., 1966. *The Tweed Ring*. New York.

15   *The New York Times*, March 7, 1871, p. 5.

16   Wingate, C. F., 1875. An episode in municipal government. 3. The Ring Charter. *North American Review*, 111, pp. 113–155. The demise of Central Park is reported on pp. 120–123.

17   Board of Commissioners of the Department of Public Parks, 1871. *First Annual Report*, pp. 18–20. This was the first (and only) Tweed Ring *Report*.

18   Hawkins had apparently built yet another full-scale *Hadrosaurus* before his soujourn at Princeton, presumably at the time when his Central Park studio was intact. The Director of the Museum of Natural History, Dr Donald Baird, writes: 'He had earlier erected in our museum a plaster skeleton of *Hadrosaurus foulkii* (the first dinosaur skeleton to be set up in any college museum), and now, between 1875 and 1877, he produced a series of 17 large paintings of prehistoric scenes. These paintings stemmed directly from his research for the Crystal Palace and Palaeozoic Museum and they included all the same animals, plus others. For thirty years they hung from the balcony railing in Nassau Hall, in the room which served in succession as the college chapel, meeting place for the Continental Congress, library, museum, and (now) faculty room.'
  'Fifteen of the original seventeen paintings still hang in Guyot Hall in various states of preservation. Some are finished and signed, some unsigned, some unfinished.' Pers. comm. Nov. 2, 1974.

19   Schuchert, C., and LeVene, C. M., 1940. *O. C. Marsh: Pioneer in Palaeontology*, New Haven, p. 385.

20   Huxley, T. H., 1868. On the animals which are most nearly intermediate between birds and reptiles. *Geol. Mag.*, 5, pp. 357–365.

21   Hitchcock, E., 1848. An attempt to discriminate and describe the animals that made the fossil footmarks of the United States, and especially of New England. *Memoirs of the American Academy of Arts and Sciences*, New Series 3, pp. 129–256. See pp. 250–251.

22   Hitchcock, E., 1836. Ornithichology. – Description of the foot marks of birds, (Ornithichnites) on new Red Sandstone in Massachusetts. *Am. Jour. Sci.*, 29, pp. 307–340. See p. 313.

23   Dollo, L., 1882. Première note sur les dinosauriens de Bernissart. *Bulletin de Musée Royal d'Histoire Naturelle de Belgique*, 1, pp. 161–180. This was the first in a long series of such notes. Dupont, E., 1870. Sur la découverte d'ossements d'Iguanodon, de poissons et de végétaux dans la Fosse Sainte-Barbe du Charbonnage de Bernissart. *Bulletin de l'Académie Royale, des Lettres et des Beaux Arts de Belgique*, Second Series, 46, pp. 387–408.

24   Brown, B., 1938. The mystery dinosaur. *Natural History*, 41, pp. 190–202, 235.

25   Galton, P. M., 1974. The ornithischian dinosaur *Hypsilophodon* from the Wealden of the Isle of Wight. *Bull. Br. Mus. nat. Hist. (Geol.)*, 25 (1), pp. 1–152.

26   Osborn, H. F., 1911. A dinosaur mummy. *Amer. Mus. Jour.*, 11, pp. 7–11.

27   Lambe, L. M., 1914. On a new genus and species of carnivorous dinosaur from the Belly River formation of Alberta, with a description of *Stephanosaurus marginatus* from the same horizon. *Ottawa Nat.*, 28, pp. 17–20.

28   Lambe, L. M., 1920. The hadrosaur *Edmontosaurus* from the Upper Cretaceous of Alberta. *Canada Dept. Mines, Geol. Surv.*, Memoir 120, pp. 1–79.

29   For some personal reminiscences and an account of the more colourful episodes in Nopcsa's life see: Edinger, T., 1955. Personalities in paleontology: Nopcsa. Society of Vertebrate Paleontology; *News Bulletin*, No. 43, pp. 35–39. See also Colbert, E. H., 1971. *Men and Dinosaurs*. Pelican, pp. 128 *et seq*.

30   Nopcsa, F., 1929. Sexual differences in ornithopodous dinosaurs. *Palaeobiologica*, 2, pp. 187–201. Anatomical differences in fossil animals of the same type has often been referred to sexual dimorphism. Pelycosaurs like *Dimetrodon* with a sail along the ridge of its back (supported by elongated neural spines) have been thought the male expression of the sail-less *Sphenacodon*. Colbert points out that the 'male' *Dimetrodon* is found in Texas and the 'female' *Sphenacodon* in New Mexico. These two localities were separated by a seaway in Early Permian times, when these reptiles flourished. An 'obstacle to discourage the most persistently amorous of these reptiles', adds Colbert. Colbert, E. H., 1969. *Evolution of the Vertebrates*. New York, p. 131.

31   Wilfarth, M., 1938. Gab es russeltragande dinosaurier? *Zeitschr. Deutsch. Geol. Ges.*, Band 40, Heft 2, pp. 88–100. For the reply see: Sternberg, C. M., 1939. Were there proboscis-bearing dinosaurs? *Ann. Mag. Nat. Hist.*, Series 2 (3), pp. 556–560.

32   Ostrom, J. H., 1962. The cranial crests of hadrosauian dinosaurs. *Postilla*, No. 62, pp. 1–29.

33   Ostrom, J. H., 1964. A reconsideration of the paleocology of hadrosaurian dinosaurs. *Am. Jour Sci.*, 262, pp. 975–997.

34   Halstead, L. B., 1969. *The Pattern of Vertebrate Evolution*. Edinburgh p. 124.

35   'The front legs were absurdly small,' says Romer in his standard manual on vertebrate palaeontology, 'and seem to have been practically useless; they were too short to reach the mouth and seemingly too weak to have been of any assistance in seizing or rending the prey.' Romer, A. S., 1966. *Vertebrate Paleontology*. 3rd ed., Chicago and London, p. 153.

36   Lambe, L. M., 1917. The Cretaceous theropodous dinosaur *Gorgosaurus*. *Canada Dept. Mines, Geol. Surv.*, Memoir 100, pp. 1–84. See p. 64.

37 Newman, B. H., 1970. Stance and gait in the flesh-eating dinosaur *Tyrannosaurus*. *Biol. J. Linn. Soc.*, pp. 119–123. Newman's radical departure from orthodox restorations will undoubtedly be contested. However, his views on the standing and walking pose of *Tyrannosaurus* do concur extremely well with Lambe's anlysis of *Gorgosauus*.

38 Ostrom, J. H., 1969. Terrestrial vertebrates as indicators of Mesozoic climates. *Proc. North Amer. Paleontol. Conv.*, pp. 347–376. The significance of the correlation between fully erect posture, endothermy, and the ability to maintain a fast metabolism was first recognised by Ostrom, although the theme was developed by Robert T. Bakker. Bakker, R. T., 1971. Dinosaur physiology and the origin of mammals. *Evolution*, 27, pp. 636–658. The correlation was challenged in 1973 on the basis of the apparent lack of any causal connection. See Feduccia, A., 1973. Dinosaurs as reptiles. *Evolution*, 27, pp. 166–170. Bennett, A. F., and Dalzell, B., 1973. Dinosaur physiology: a critique. Ibid., pp. 170–174. Ostrom replied that as all modern terrestrial vertebrates with an erect posture were endothermic, and since dinosaurs likewise possessed an erect posture, then there is a high degree of probability that dinosaurs too were endothermic. Whether the causal linkage has been identified yet or not, answered Ostrom, the correlation between posture and metabolism still stands. Ostrom, J. H., 1974. Reply to 'Dinosaurs as reptiles'. *Evolution*, 28, pp. 491–493. See also Bakker's and Peter Dodson's answers to the charges, and their rejoinders by Bennett and Feduccia in the same volume.

39 Russell, D. A., 1973. The environments of Canadian dinosaurs. *Canad. geog. J.*, 87 (1), pp. 4–11.

3. The race is to the swift, the battle to the strong

1 Ricqlès, A. de, 1969. L'histologie osseuse envisagée comme indicateur de la physiologie thermique chez les tétrapodes fossiles. *Comptes Rendus Acad. Sci.*, Paris, 268D, pp. 782–785. This paper was only a note of Ricqlès' inferences; he reported his full findings concerning dinosaur bone the previous year, but drew few inferences at the time. Ricqlès, A. de, 1968. Recherches paléohistologie sur les os longs des tétrapodes. I. Origine du tissu osseux plexiforme des dinosauriens sauropodes. *Ann. Paléontol.(Vertébrés)*, 54, pp. 133–145.

2 Lambe, L. M., 1917. The Cretaceous theropodous dinosaur *Gorgosaurus*. *Canada, Depart. Mines; Geol. Surv.*, Memoir 100, Ottawa, pp. 1–84.

3 Matthew, W. D., 1915. *Dinosaurs*. New York, American Museum of Natural history: Handbook No. 5, pp. 40–41.

4 Osborn, H. F., 1905. *Tyrannosaurus* and other Cretaceous carnivorous dinosaurs. *Bull. Amer. Mus. Nat. Hist.*, 21, pp. 259–265.

5 Osborn, H. F., 1913. *Tyrannosaurus*, restoration and model of the skeleton. *Bull. Amer. Mus. Nat. Hist.*, 32, pp. 91–92.

6 Brown, B., 1915. *Tyrannosaurus*, a Cretaceous carnivorous dinosaur. The largest flesh-eater that ever lived. *Scientific American*, 113, pp. 322–323.

7 Ostrom, in his description of the extraordinary carnivore *Deinonychus*, states that 'The multiple remains of [*Deinonychus*] suggest that [it] may have been gregarious and hunted in packs. Ostrom, J. H., 1969. Osteology of *Deinonychus antirrhopus*, an unusual theropod from the Lower Cretaceous of Montana. *Bull. Peabody Museum of Natural History, Yale University*, 30, pp. 1–165. See p. 144.

8 Matthew, W. D., and Brown, B., 1923. Preliminary note of skeletons and skulls of Deinodontidae from the Cretaceous of Alberta. *Amer. Mus. Novit.*, 89, pp. 1–10.

9 Matthew, op. cit. (3), p. 45.

10 Swinton, W. E., 1934. *The Dinosaurs: A Short History of a Great Group of Extinct Reptiles*. London, p. 70.

11 Bird, R. T., 1954. We captured a 'live' brontosaur. *Nat. geogr. Mag.*, 105 (5), pp. 707–722.

12 Lull, R. S., 1917. *Organic Evolution*. New York, pp. 414–415.

13 Kielan-Jaworowska, Z., and Barsbold, R., 1972. Results of the Polish–Mongolian palaeontological expedition – part IV. *Palaeontol. Polonica*, 27, pp. 5–13.

14 Bakker, R. T., 1972. Anatomical and ecological evidence of endothermy in dinosaurs. *Nature*, 238, pp. 81–85. See p. 85 (my emphasis). It must be pointed out that in some instances, especially in Mongolia, finds consist almost entirely of carnivores. In these cases, it is possible that quicksand or tar pits snared the creatures. It is known from the Californian Pleistocene tar pits, in which many killers like sabre-tooth 'tigers' lie embedded, that the finds are not representative of the communities as a whole. Stranded creatures attract an unnatural proportion of carnivores to the scene.

15 Bakker, R. T., 1968. The superiority of dinosaurs. *Discovery* (New Haven), 3 (2), pp. 11–22. See p. 21.

16 Beecher, C. E., 1902. The reconstruction of a Cretaceous dinosaur, *Claosaurus annectans* Marsh. *Trans. Conn. Acad. Arts and Sci.*, 11, pp. 311–324.

17 Ostrom, op. cit. (7), p. 139.

18 Galton, P. M., 1970. The posture of hadrosaurian dinosaurs. *J. Paleo.*, 44, pp. 464–473.

19 Bakker, op. cit. (14), p. 81.

20 Colbert, E. H., 1962. *Dinosaurs : Their Discovery and their World.* London, p. 87.

21 Osborn, H. F., 1916. Skeletal adaptations of *Ornitholestes, Struthiomimus, Tyrannosaurus. Bull. Amer. Mus. Nat. Hist.*, 35, pp. 733–771.

22 'Notes on the habits of *Struthiomimus* by Dr William K. Gregory.' Ibid. pp. 758–760.

23 This restoration by Heilmann caused Swinton to comment : 'The generic name means ostrich-mimic, and unquestionably the structure of the skeleton in many ways is closely alike to that of the struthious birds we know so well in life today. These features, and perhaps a too literal translation of the name, have caused some artists, notably G. Heilmann (1926), to produce a reconstruction so ostrich-like in appearance that, despite the presence of fore limbs and a long tail, the first impression is hardly reptilian . . .' But then it is not surprising that the struthiomimid looks so unreptilian for, as Swinton continues, 'it must be remembered that not very many truly reptilian features dominate the appearance of [its] skeletal structure'. Swinton, op. cit. (10), p. 77.

24 Heilmann, G., 1925. *The Origin of Birds.* London, p. 183, Fig. 132, reproduced here as fig. 29. This book is a mine of superbly drawn illustrations of *running* dinosaurs.

25 Bakker, op. cit. (14), p. 81.

26 Gregory, W. K., 1912. Notes on the principles of quadrupedal locomotion and on the mechanism of the limbs in hoofed animals. *Annals N. Y. Acad. Sci.*, 22, pp. 267–294. See plate 34.

27 Figures taken from : Galton, P. M., 1974. The ornithischian dinosaur *Hypsilophodon* from the Wealden of the Isle of Wight. *Bull. Br. Mus. Nat. Hist. (Geol.)*, 25 (1), pp. 1–152. See Table V, p. 139.

28 Russell, D. A., 1972. Ostrich dinosaurs from the Late Cretaceous of Western Canada. *Can. J. Earth Sci.*, 9, pp. 375–402. Russell was able to show that these dinosaurs had large brains (in *Dromiceiomimus* the brain exceeded an ostrich's in size). The eyes were huge, the area of retina was larger than in any terrestrial vertebrate, and vision extremely acute. Parental care of the young was not beyond the mental capacity of these dinosaurs and Russell speculates on this and other intriguing problems.

29 In 1912 Gregory also noted the 'connection between the mode of locomotion and the length of the middle metatarsal as compared with the femur'. Gregory, op. cit. (26), pp. 284 *et seq.* This ratio is even more extreme than the T/F and its expresses the degree to which the creature has risen on to its toes as an aid to fast running. For example, in heavy graviportal mammals the MT/F ranges from 0.10 to 0.20 (0.13 in the elephant) indicating that the metatarsals of the toes have barely elongated at all ; in rhinos it is 0.37 ; in early subcursorial horses 0.50–0.60 ; whilst in the modern horse the metatarsals have elongated to such an extent that the animal walks on the tips of its toes, and has a MT/F ratio of 0.78. For struthiomimids the ratio is 0.77, in *Hypsilophodon* 0.62, and both dinosaurs likewise run on their toes. In the giant sauropod dinosaurs the figure has dropped to 0.13 (*Apatosaurus*). These figures tell the same story as the shin : thigh ratios. Figures for dinosaurs taken from Galton, op. cit. (18), pp. 467–468 ; and Galton, op. cit. (27).

4. The Dark Ages

1 Newton, E. T., 1894. Reptiles from the Elgin Sandstone. – Description of two new genera. *Phil. Trans. Roy. Soc.*, 185 B, Part 1, pp. 573–604.

2 Boulenger, C. A., 1904. On reptilian remains from the Trias of Elgin. *Phil. Trans. Roy. Soc.*, 196 B, pp. 175–189.

3 Broom, R., 1913. On the South African pseudosuchian *Euparkeria* and allied genera. *Proc. Zool. Soc. Lond.*, (2), pp. 619–633.

4 Ibid. p. 630.

5 Colbert, E. H., 1962. *Dinosaurs : Their Discovery and their World.* London, pp. 66–67.

6 Walker has placed it with the dinosaurian flesh-eaters, but this is hotly disputed. Walker, A. D., 1964. Triassic reptiles from the Elgin area: *Ornithosuchus* and the origin of the carnosaurs. *Phil. Trans. Roy. Soc.*, 248 B, pp. 53–134. Although Romer followed suit and placed *Ornithosuchus* among the carnosaurs in his *Vertebrate Paleontology* (1966), he later had second thoughts and demoted it to the Pseudosuchia. Romer, A. S., 1972. The Chañares (Argentina) Triassic reptile fauna. XIII. An early ornithosuchid pseudosuchian, *Gracilisuchus stipanicicorum*, gen. et sp. nov. *Breviora*, 389, pp. 1–24. See p. 23.

7 Ewer, R. F., 1965. The anatomy of the thecodont reptile *Euparkeria capensis* Broom. *Phil. Trans. Roy. Soc.*, 428 B, pp. 379–435.

8 Romer, A. S., 1966. *Vertebrate Paleontology.* 3rd. ed., Chicago and London, p. 137.

9 Ewer, op. cit. (7).

10 Reig, O., 1970. The Proterosuchia and the early evolution of the archosaurs; an essay about the origin of a major taxon. *Bull. Mus. Comp. Zool.*, Harvard, 139 (5), pp. 229–292.

11 Olson, E. H., 1971. *Vertebrate Paleozoology.* New York, pp. 664 and 666.

12 Watson, D. M. S., 1951. *Paleontology and Modern Biology.* New Haven, pp. 38–78.

13 Reig, op. cit. (10), pp. 259–262.

14 Romer alludes to this point. 'It has been suggested that the initiation of these changes [he is referring to the development of long hind limbs and counterbalancing tail in bipeds] was associated with improvements for swimming in archosaur ancestors of amphibious habits. If so, this was a fortunate "preadaptation".' Romer, op. cit. (8), p. 136. Charig states this far more specifically: 'Could it not be that the earliest archosaurs were amphibious – like mesosaurs and otters – using their tail for propulsion in the water and their hind limbs for paddling, steering or pushing on the bottom? The greater importance of the hind limb in aquatic locomotion is shown by the increased limb disparity in marine crocodiles and by the fact that in modern crocodiles it is only the hind foot that is webbed.' Charig, A. G., 1966. Stance and gait in the archosaur reptiles. *Advmt. Sci.*, Lond., 22, p. 537 (Abstr.). When startled, crocodiles will occasionally rear up and run for short spurts on their hind legs.

15 Romer, A. S., 1971. The Chañares (Argentina) Triassic reptile fauna. X. Two new but incompletely known long-limbed pseudosuchians. *Breviora*, 378, pp. 1–10.

16 The key innovation triggering the emergence of dinosaurs was probably bipedalism (and thus the vertical limb posture) matched to a sustained fast-metabolising physiology. Charig believes that it was the upright stance alone that conferred a special advantage upon dinosaurs. Thus he believes that erect quadrupeds gave rise to the bipeds as well as to persistently four-footed dinosaurs. Since agile long-striding bipedal pseudosuchians made their appearance as early as the Lower-Middle Trias, and were obviously eminently successful, even evolving forms (such as *Ornithosuchus*) paralleling the bipedal flesh-eating dinosaurs, it is difficult to believe that they did not already hold the key innovation later inherited and put to even better effect by dinosaurs. Charig, A. G., 1972. The evolution of the archosaur pelvis and hindlimb: an explanation in functional terms. In Joysey, K. A., and Kemp, T. S. (eds). *Studies in Vertebrate Evolution.* Edinburgh, pp. 121–155.

17 Bakker, R. T., and Galton, P. M., 1974. Dinosaur monophyly and a new Class of vertebrates. *Nature*, 248, pp. 168–172.

18 Seeley, H. G., 1887. The classification of the Dinosauria. *Report of the British Association for the Advancement of Science*, Manchester, 57, pp. 698–699.

19 Brink, A. S., 1956. Speculations on some advanced mammalian characteristics in the higher mammal-like reptiles. *Palaeont. Africana*, 4, pp. 77–95. Armand de Ricqlès completely endorses the view that therapsids were endotherms. Ricqlès has shown from his histological studies that therapsid bone, being highly vascular and with numerous Haversian canals, can be classed with mammal bone, in contradistinction to the bones of fossil and living reptiles and amphibians. Ricqlès, A. de, 1974. Recherches paléohistologiques sur les os longs des tétrapodes. IV. Éothériodontes et pélycosaures. *Ann. Paleontol. (Vertébrés)*, 60, pp. 3–39. See also Ricqlès, A. de, 1969. L'histologie osseuse envisagée comme indicateur de la physiologie thermique chez les tétrapodes fossiles. *Comptes Rendus Acad. Sci.*, Paris, 268D, pp. 782–785. Ricqlès, A. de, 1969. Recherches paléohistologiques sur les os longs des tétrapodes. II. Quelques observations sur la structure des os longs des thériodontes. *Ann. Paleontol. (Vertébrés)*, 55, pp. 1–52.

20 Watson, D. M. S., 1931. On the skeleton of a bauriamorph reptile. *Proc. Zool. Soc. Lond.*, pp. 1163–1205. See p. 1169.

21 Bakker, R. T., 1971. Dinosaur physiology and the origin of mammals. *Evolution*, 25, pp. 636–658.

22 Crompton, A. W., 1968. The enigma of the origin of mammals. *Optima* (Anglo American Corp.), 18, pp. 137–151. See p. 144.

23 Jenkins, F. A., Jr, 1971. The postcranial skeleton of African cynodonts: problems in the evolution of mammalian postcranial anatomy. *Yale Peabody Mus. Bulletin*, 36, pp. 1–216. Bakker, op. cit. (21), would have them sprawling even more than Jenkins allows.

24 Crompton, op. cit. (22), p. 143.

25 Robinson, P. L., 1969. A problem of faunal replacement on Permo-Triassic continents. *Palaeontology*, 14, pp. 131–153. See pp. 135–137.

26 Kermack, K. A., and Mussett, F., 1959. The first mammals. *Discovery* (London), 20, pp. 144–151. Kermack, D. M., Kermack, K. A., and Mussett, F., 1968. The Welsh pantothers *Kuehneotherium praecursoris*. *J. Linn. Soc. (Zool.)*, 47, pp. 407–423.

27 Hopson, J. A., 1971. Postcanine replacement in the gomphodont cynodont *Diademodon*. *In* Kermack, D. M., and Kermack, K. A. (eds). *Early Mammals*. Zool. *J. Linn. Soc. Lond.*, 50, Suppl. 1, pages 1–21. See pp. 18–20. Also Brink, op. cit. (19).

28 Gill, P. G., 1974. Resorption of premolars in the early mammal *Kuehneotherium praecursoris*. *Archs. oral Biol.*, 19, pp. 327–328.

29 Kermack, K. A., Mussett, F., and Rigney, H. W., 1973. The lower jaw of *Morganucodon*. *Zool. J. Linn. Soc. Lond.*, 53, pp. 87–175. An account of the quarrying of the Welsh fissure material is also given in this paper.

30 Crompton, A. W., 1968. In search of the 'insignificant'. *Discovery* (New Haven), 3 (2), pp. 23–32.

31 Colbert, E. H., and Russell, D. A., 1969. The small Cretaceous dinosaur *Dromaeosaurus*. *Amer. Mus. Novit.*, 2380, pp. 145–162.

32 Russell, D. A., 1969. A new specimen of *Stenonychosaurus* from the Oldman Formation (Cretaceous) of Alberta. *Can. J. Earth Sci.*, 6, pp. 595–612.

33 Colbert, op. cit. (5), p. 199.

34 Romer, A. S., 1962. *Man and the Vertebrates*. Penguin, vol. 1, pp. 95–96.

35 Cobb, S., and Edinger, T., 1962. The brain of the emu (*Dromaeus* novaehollandiae, Lath.). *Breviora*, 170, pp. 1–18. See p. 5.

36 Russell, D. A., 1972. Ostrich dinosaurs from the Late Cretaceous of Western Canada. *Can. J. Earth Sci.*, 9, pp. 375–402.

37 Russell, op. cit. (32).

5. The stranding of the Titans

1 Marsh, O. C., 1877. Notice of a new and gigantic dinosaur. *Am. Jour. Sci.*, 14, pp. 87–88.

2 Buckland, W., 1837. *Geology and Mineralogy*. London, p. 115.

3 Owen, W., 1841. Report on British fossil reptiles, Part 2. *Report of the Eleventh Meeting of the British Association for the Advancement of Science*, London, see pp. 94–103.

4 Owen, R., 1841. A description of a portion of the skeleton of the Cetiosaurus, a gigantic extinct Saurian Reptile occurring in the Oolite formations of different portions of England. *Proc. Geol. Soc.* London, Vol. 3, Part 2, No. 80, pp. 457–462.

5 Owen, R., 1874–1889. *Monograph of the Fossil Reptilia of the Mesozoic Formations*. Palaeontographical Society, p. 412.

6 Owen, R., 1847. Report on the archetype and homologues of the vertebrate skeleton. (Published separately; it originally appeared in the *Report of the British Association for the Advancement of Science* for 1846), London, pp. 169–340. Owen later adopts a stronger evolutionary position, for example in his *Monograph*, op. cit. (5), pp. 87–93.

7 Phillips, J., 1870. *Cetiosaurus*. *Athenaeum*, No. 2214, April 2, p. 454. See also Phillips, J., 1871. *Geology of Oxford and the Valley of the Thames*. Oxford, pp. 254–294.

8 Owen, op. cit. (5), p. 564.

9 Huxley, T. H., 1870. Further evidence of the affinity between the dinosaurian reptiles and birds. *Proc. Geol. Soc.* London, 26, p. 12. Also Huxley, T. H., 1870. On the classification of the Dinosauria, with observations on the Dinosauria of the Trias. Ibid., p. 35.

10 Seeley, H. G., 1874. On the base of a large lacertilian cranium from the Potton Sands, presumably dinosaurian. *Quart. Jour. Geol. Soc.*, 30, pp. 690–692.

11 Owen, op. cit. (5), p. 593.

12 This account of the Marsh-Cope controversy is based on Schuchert, C., and LeVene, C. M., 1940. *O. C. Marsh, Pioneer in Paleontology*. New Haven, pp. 189 *et seq*; Ostrom, J. H., and McIntosh, J. S., 1966. *Marsh's Dinosaurs*. New Haven and London, pp. 2–47; and Williston, S. W., 1915. The first discovery of dinosaurs in the West. *In* Matthews, W. D., 1915. *Dinosaurs*. New York, A.M.N.H., pp. 124–131.

13 Marsh, O. C., 1877. A new order of extinct Reptilia (Stegosauria) from the Jurassic of the Rocky Mountains. *Am. Jour. Sci.*, 14, pp. 513–514.

14 Cope, E. D., 1877. On the Vertebrata of the Dakota Epoch of Colorado. *Proc. Amer. Philos. Soc.*, 17, pp. 233–247.

15 Marsh, O. C., 1878. Principal characters of American Jurassic dinosaurs. *Am. Jour. Sci.*, 16, pp. 411–416.

16 Marsh, O. C., 1883. Principal characters of American Jurassic dinosaurs, Part VI, Restoration of Brontosaurus. *Am. Jour. Sci.*, 26, pp. 81–85.

17 Ballou, W. H., 1897. Strange creatures of the past. *The Century Illustrated Monthly Magazine*, New York, 55, pp. 15–23. Cope drew another sketch of *Amphicoelias* standing on a lake floor. This is reproduced in Osborn, H. F., 1931. *Cope: Master Naturalist*. Princeton Univ. Press, opposite p. 302. Osborn's date for this sketch (1878) seems too late, since the sauropods are depicted with diminutive front limbs and were obviously thought to be bipeds. In 1877 Cope was already aware of the massive front limbs of *Camarasaurus*.

18 Osborn, H. F., 1898. Additional characters of the great herbivorous dinosaur *Camarasaurus*. *Bull. Am. Mus. Nat. Hist.*, 10, pp. 219–233.

19 See Wall, J. F., 1970. *Andrew Carnegie*. New York, esp. Ch. 11 and 12. Carnegie, A., 1965. *The Gospel of Wealth, and Other Timely Essays*. Harvard Univ. Press, Cambridge, Mass. [A reprint of the first, 1900, edition.]

20 Hatcher's preliminary restoration appeared in: Hatcher, J. B., 1901. *Diplodocus* Marsh, its osteology, taxonomy, and probable habits, with a restoration of the skeleton. *Mem. Carnegie Mus.*, 1, pp. 1–63. Holland's amended restoration was published in: Holland, W. J., 1905. The osteology of *Diplodocus* Marsh with a special reference to the restoration of the skeleton of *Diplodocus carnegiei* Hatcher presented by Mr Andrew Carnegie to the British Museum, May 12, 1905. *Mem. Carnegie Mus.*, 2, pp. 225–264.

21 Anonymous [Holland, W. J.], 1905. The presentation of a reproduction of *Diplodocus carnegiei* to the Trustees of the British Museum. *Ann. Carnegie Mus.*, 3, pp. 443–452.

22 Frohawk, F. W., 1905. The attitude of *Diplodocus carnegiei*. *The Field*, 106, p. 388. For the reply vindicating the Americans for their choice of pose see *The Field*, 106, p. 466.

23 Hay, O. P., 1908. On the habits and the pose of the sauropodous dinosaurs, especially of *Diplodocus*, *Am. Nat.*, 42, pp. 672–681.

24 Hay, O. P., 1910. On the manner of locomotion of the dinosaurs especially *Diplodocus*, with remarks on the origin of birds, *Proc. Wash. Acad. Sci.*, 12, pp. 1–25. Plate 1.

25 Holland, W. J., 1910. A review of some recent criticisms of the restorations of sauropod dinosaurs existing in the museums of the United States, with special reference to that of *Diplodocus carnegiei* in the Carnegie Museum. *Am. Nat.*, 44, pp. 259–283.

26 Bird, R. T., 1954. We captured a 'live' brontosaur. *National Geographic*, 105, pp. 707–722.

27 Matthew, W. D., 1910. The pose of sauropodous dinosaurs. *Am. Nat.*, 44, pp. 547–560.

28 Hatcher, op. cit. (19), p. 60.

29 Holland, W. J., 1924. The skull of *Diplodocus*. *Mem. Carnegie Mus.*, 9, pp. 379–403.

30 Osborn, H. F., 1915. The dinosaurs of the Bone Cabin quarry. *In* Matthew, W. D., 1915. *Dinosaurs*. New York, A.M.N.H., pp. 131–152.

31 Ibid. p. 141.

32 Bird, R. T., 1944. Did Brontosaurus ever walk on land? *Natural History*, 53, No. 2, pp. 60–67.

33 Kermack, K. A., 1951. A note on the habits of the sauropods. *Ann. Mag. Nat. Hist.*, 12 (4), pp. 830–832. For Stigler's paper, correlating the depth of immersion, water pressure, and the ability (or inability) of human subjects to breathe, see: Stigler, R., 1911. Di Kraft unserer Inspirationsmuskulatur. *Pflüger's Archiv für Physiologie*, 139, pp. 234–254. In fact, Oliver P. Hay in 1910, op. cit. (24), p. 25, may have anticipated Kermack's line of reasoning. Hay wrote:

> In the paper published by Mr Ballou . . . there is a figure which represents a group of four individuals of *Amphicoelias latus* . . . These animals are shown as walking about on the bottom of a river, feeding on the vegetation there and rising on their hind legs to reach the air . . . It would seem to have been hardly more possible for *Diplodocus* to walk about immersed in water than it would be for a man to do the same.

34 Colbert. E. H., 1952. Breathing habits of the sauropod dinosaurs. *Ann. Mag. Nat. Hist.*, 12 (25), pp. 708–710.

35 Kermack in conversation with the author on October 18, 1974.

36 Swinton writes in 1967 that the big sauropods 'most probably lived in the waters of estuaries or lakes where the buoyancy would lessen the cumbersome body weight' and that 'In *Diplodocus* . . . the nostril was an opening on top of the skull. In deepish water, surrounded by an abundance of food and free from the attacks of land carnivores, the Sauropods were in satisfactory surroundings'. Swinton, W. E., 1967. *Dinosaurs*. B.M.(N.H.), London, pp. 17, 19. The most frequently used standard work in vertebrate palaeontology is Romer's *Vertebrate Paleontology*. Here we find the same views expressed: 'The position of these [nasal] organs suggests an amphibious mode of life for the sauropods; the animal could breathe and look about with only the top of the head exposed above the water.' Romer, A. S., 1966. *Vertebrate Paleontology*. Chicago and London, p. 155.

37 Bakker, R. T., 1971. Ecology of the brontosaurs. *Nature*, 229, pp. 172–174. Bakker's novel ideas concerning sauropods were first published some three years earlier: Bakker, R. T., 1968. The superiority of dinosaurs. *Discovery* (New Haven), 3 (2), pp. 11–22.

38 News and View, 1971. Changing dinosaurs – but not in mid-stream. *Nature*, 229, p. 153.

39 Bakker, 1971, op. cit. (37), p. 174.

40 Bird, op. cit. (32).

41 Ostrom, J. H., 1972. Were some dinosaurs gregarious? *Palaeogeogr., Palaeoclimatol., Palaeoecol.*, 11, pp. 287–301.

42 Richmond, N. D., 1965. Perhaps juvenile dinosaurs were always scarce. *J. Palaeo.*, 39, pp. 503–505.

43 Sikes, S. K., 1971. *The Natural History of the African Elephant*. London, pp. 102–107.

44 Brown, B., 1941. The last dinosaurs. *Natural History*, 48, pp. 290–295.

6. A Griffin rescues evolution

1 Mivart, St. G., 1871. *The Genesis of Species*. London. See Chapter IV and especially pp. 106–107.

2 Meyer, H. von, 1862. On the *Archaeopteryx lithographica* from the lithographic slate of Solnhofen. *Ann. Mag. nat. Hist.*, 9, pp. 366–370.

3 Wagner, A., 1862. On a new fossil reptile supposed to be furnished with feathers. *Ann. Mag. nat. Hist.*, 9, pp. 261–267.

4 Quoted by de Beer, G., 1954. *Archaeopteryx lithographica*: A study based upon the British Museum specimen. British Museum (Natural History), p. 2.

5 A delightfully bad pun borrowed from Herbert Wendt, 1970. *Before the Deluge*. Paladin paperback, p. 240.

6 Huxley, T. H., 1868. On the animals which are most nearly intermediate between birds and reptiles. *Ann. Mag. nat. Hist.*, (4), 2, pp. 66–75.

7 Vogt, C., 1880. *Archaeopteryx macrura*, an intermediate form between birds and reptiles. *Ibis*, 4, pp. 434–456.

8 Heilmann, G., 1925. *The Origin of Birds*. London, p. 165.

9 Osborn, H. F., 1916. Skeletal adaptations of *Ornitholestes, Struthiomimus, Tyrannosaurus*. *Bull. Am. Mus. Nat. Hist.*, 35, pp. 733–771.

10 Cope, E. D., 1867. [An account of the extinct reptiles which approached the birds.] *Proc. Acad. Nat. Sci. Philadelphia*, pp. 234–5.

11 Huxley, op. cit. (6), and 1870. On the classification of the Dinosauria, with observations on the dinosaurs of the Trias. *Q. Jour. Geol. Soc.*, 26, pp. 32–51.

12 Simpson adopted this expedient to discredit Lowe (who believed that non-flying birds arose directly from dinosaurs). By making coelurosaurs and birds merely convergent, Simpson removed the dinosaur ancestor and so made it more plausible that *all* birds were derived from *Archaeopteryx* (which Lowe dismissed as a sterile offshoot). Simpson, G. G., 1946. Fossil penguins. *Bull. Am. Mus. Nat. Hist.*, New York, 87, pp. 1–99. For the polemic against Lowe, see 'A note on *Archaeopteryx* and *Archaeornis*', pp. 92–95.

13 Heilmann was forced into the position of having to request a letter of recommendation from Arthur Smith Woodward of the British Museum. Letter to Woodward dated October 10, 1925, accompanied by a copy of the book: Smith Woodward collection housed in the D. M. S. Watson Natural Sciences Library at University College London.

14 De Beer, op. cit. (4), see pp. 42–50.

15 Walker, A. D., 1972. New light on the origin of birds and crocodiles. *Nature*, 237, pp. 257–263.

16 Ostrom, J. H., 1973. The ancestry of birds. *Nature*, 242, p. 136.

17 In fact, this dislocation is quite apparent in the text-figure of the pelvic girdle of the Berlin specimen in de Beer's monograph, op. cit. (4), text-fig. 7, p. 29.

18 Bakker, R. T., and Galton, P. M., 1974. Dinosaur monophyly and a new Class of vertebrates. *Nature*, 248, pp. 168–172.

19 *The Sunday Times*, March 17, 1974, p. 13.

20 Ibid.

21 Ostrom, J. H., 1974. *Archaeopteryx* and the origin of flight. *Q. Rev. Biol.*, 49, pp. 27–47.

22 Bakker and Galton, op. cit. (18).

23 Walker, op. (15), points out that the supracoracoideus wing elevator muscle was not functional in *Archaeopteryx*, and Ostrom (pers comm. March 4, 1975) notes the reduction or complete loss of the usual elevators, the deltoids, as evidenced by the narrow scapula.

24 De Beer, op. cit. (4), p. 24.

25 Halstead, L. B., 1969. *The Pattern of Vertebrate Evolution*. Edinburgh, p. 143.

26 Heptonstall, W. B., 1970. Quantitative assessment of the flight of *Archaeopteryx*. *Nature*, 228, pp. 185–186. Heptonstall's weight estimate of 500 grams is based on that of the similar sized pigeon (400 grams) plus an additional amount to allow for the lack of pneumatic bones, etc.

27 Yalden, D. W., 1971. Flying ability of *Archaeopteryx*. *Nature*, 231, p. 127. Bramwell, C., 1971. *Nature*, 231, p. 128. See Heptonstall's reply to both critics, ibid. Yalden's estimate of the weight of *Archaeopteryx* is 200 grams, i.e., that of a light but similar sized bird today. This seems far too light for a scarcely modified coelurosaurian dinosaur.

28 Mayr, E., 1960. The emergence of evolutionary novelties. *In* S. Tax (ed.). *The Evolution of Life*. Chicago. Bock, W. J., 1969. The Origin and radiation of birds. *Ann· N. Y. Acad. Sci.*, 167, pp. 147–155.

29 Ostrom, op. cit. (21), pp. 43–44.

30 Bock, op. cit. (28).

31 De Beer, op. cit. (4), p. 38.

32 Ostrom, J. H., 1974. Reply to 'Dinosaurs as reptiles'. *Evolution*, 28, pp. 491–493. *Compsognathus*, although larger than *Archaeopteryx*, was itself small and likewise found in the Solnhofen sediments. It is conceivable that feathers were present on this coelurosaur. According to Ostrom (pers. comm. March 4, 1975) the *Compsognathus* slab has been polished smooth around the fossil, so any faint feather impressions would have been obliterated. *Compsognathus* was discovered *before* the first (London) *Archaeopteryx* specimen, so at the time of its appearance no one had become alerted to the possibility of feathers. Feather impressions in the last three *Archaeopteryx* specimens were so faint that the creatures were not recognised as *Archaeopteryx* at first. As a final point to emphasise the similarity between the small coelurosaurs and *Archaeopteryx*, it might be noted that the fifth *Archaeopteryx* specimen, lacking distinct feather impressions, was long thought to have been *Compsognathus*.

33 Jerison, H. J., 1968. Brain evolution and *Archaeopteryx*. *Nature*, 219, pp. 1381–2.

34 Ostrom, op. cit. (21).
35 Nopcsa, Baron F., 1907. Ideas on the origin of flight. *Proc. Zool. Soc.* London, pp. 223–236. See especially pp. 234–236.
36 Duerden, J. E., 1920. Inheritance of callosities in the ostrich. *Am. Nat.*, 54, pp. 289–312.
37 For a modern Darwinian explanation of the way Lamarckism can be mimicked, see Waddington's ideas on canalisation in Maynard Smith, J., 1972. *The Theory of Evolution*. Penguin, pp. 294–299.
38 Lowe, P. R., 1926. On the callosities of the ostrich (and other Palaeognathae) in connection with the inheritance of acquired characters. *Proc. Zool. Soc.* London, pp. 667–679. Lowe, P. R., 1935. On the relationship of the Struthiones to the dinosaurs and to the rest of the avian class, with special reference to the position of *Archaeopteryx*. *Ibis*, 5, pp. 398–432. Lowe, P. R., 1944. Some additional remarks on the phylogeny of the Struthiones. *Ibis*, 86, pp. 37–42.
39 Lowe, 1935, op. cit. (37), p. 420.
40 Marsh originally found *Hesperornis* in Kansas, but more recently Dale Russell has located such plentiful remains in Arctic Canada (in places *Hesperornis* accounts for between one-third and three-quarters of all finds) that it appears as though he has stumbled upon the *Hesperornis* breeding grounds. This is made likely by the number of immature specimens occurring in the north. Russell, D. A., 1973. The environments of Canadian dinosaurs. *Canad. Geog. J.*, 87 (1), pp. 4–11.
41 Swinton, W. E., 1965. *Fossil Birds*. British Museum (Natural History), p. 41.

7. Phantoms from Hell

1 Cuvier, G., 1824. *Recherches sur les Ossimens Fossiles*. Paris, Vol. 5, part 2, pp. 379–380.
2 Seeley, H. G., 1870. *The Ornithosauria*. Cambridge, p. 7.
3 Reported in Ibid. p. 11.
4 Quoted in Lang, W. D., 1939. Mary Anning (1799–1847), and the pioneer geologists of Lyme. *Proc. Dorset nat. Hist. archaeol. Soc.*, 6, pp. 142–164.
5 Buckland, Rev. W., 1835. On the discovery of a new species of pterodactyl in the Lias at Lyme Regis. *Geol. Trans.*, London, Ser. 2, Vol. 3, pp. 217–222. See p. 218.
6 Buckland, Rev. W., 1836. *Geology and Mineralogy, Considered with reference to Natural Theology*. 1, pp. 224–225. This is volume 5 of *The Bridgewater Treatises on the Power Wisdom and Goodness of God as manifested in the Creation*, London.
7 Abel, O., 1918. Uber die Verwertung von Fossilrekonstruktionen im naturwissenschaflichen Unterricht. *Aus der Natur.*, Heft. 1, pp. 17–23. See p. 20. Also Abel, O., 1911. Die Bedeutung der fossilen Wirbeltiere für die Abstammungslehre. *Die Abstammungslehre*, pp. 198–250. See pp. 207–213.
8 Hawkins, T., 1840. *The Book of the Great Sea-Dragons, Ichthyosauri and Plesiosauri, Gedolim Taninim of Moses, Extinct Monsters of the Ancient Earth*. London, p. 26. His eccentricity was becoming more in evidence at this time. His earlier work on marine saurians had consisted of a relatively pedantic series of descriptions. He abandoned this style altogether after the 1840 work, and in 1844 produced *The Wars of Jehovah in Heaven, Earth and Hell!*
9 Newton, E. T., 1888. On the skull, brain, and auditory organ of a new species of pterosaurian (*Scaphognathus purdoni*), from the Upper Lias near Whitby, Yorkshire. *Phil. Trans. Roy. Soc.*, London, 179 B, pp. 503–537.
10 Ibid. p. 510.
11 Seeley, H. G., 1901. *Dragons of the Air*. New York (reprinted 1967, Dover paperback), p. 54.
12 Edinger, T., 1941. The brain of *Pterodactylus*. *Am. Jour. Sci.*, 239, pp. 665–682.
13 Seeley, op. cit. (11). His earlier book, *The Ornithosauria*, was only concerned with the Cambridge Greensand pterodactyls.
14 Seeley, op. cit. (11), p. 57.
15 Newman, E., 1843. Note on the pterodactyle tribe considered as marsupial bats. *The Zoologist*, 1, pp. 129–131.
16 Wanderer, K., 1908. *Rhamphorhynchus gemmingi* H. v. Meyer. Ein exemplar mit teilweise erhaltener flughaut a.d. Kgl. Mineralog.-Geol. Museum zu Dresden. *Palaeontographica*, Band 55, pp. 195–216. The 'dimpled' skin of the pterosaur is referred to on pp. 198 and 205.
17 Broili, F., 1927. Ein *Rhamphorhynchus* mit spuren von haarbedeckung. *Sitzgsb. d. math.-naturw. Abt.*, February 5, pp. 49–67.

18 Sharov, A. G., 1971. New flying reptiles from the Mesozoic deposits of Kazakhstan and Kirgizia. *Trudy. Pal. Inst. AN S.S.S.R.*, 130, pp. 104–113. (In Russian.) Mr Jack Saxon, who very kindly translated this paper for me, was also responsible for the alternative translation of *Sordes pilosus*.

19 Halstead produces a slightly different argument for parental care in pterosaurs. 'It has been suggested that pterosaurs such as this [*Pteranodon*] cared for their young, since it would not have been possible for the females to have carried eggs commensurate with the size expected. They would have had to lay very small eggs, and when newly hatched the young would have been quite unable to fend for themselves; *ergo* the parents cared for them.' Halstead, L. B., 1969. *The Pattern of Vertebrate Evolution*. Edinburgh, p. 142.

20 Seeley, op. cit. (2).

21 Seeley, op. cit. (11), p. 230.

22 Bakker, R. T., and Galton, P. M., 1974. Dinosaur monophyly and a new class of vertebrates. *Nature*, 248, p. 171.

23 Sharov, op. cit. (18).

24 Zambelli, R., 1973. *Eudimorphodon ranzii* gen. nov., sp. nov., uno pterosauro Triassico. *Istituto Lombardo (Rend. Sc.)*, B, 107, pp. 27–32. This earliest known pterosaur, found in Upper Triassic sediments in Bergamo, Italy, is extraordinary for its possession of multicuspate teeth like those of early mammals. Later pterosaurs had single cusped teeth.

25 Owen, R., 1841–1884. *A History of British Fossil Reptiles*. London, Vol. 1, pp. 245–246.

26 Ibid.

27 Marsh, O. C., 1871. Note on a new and gigantic species of Pterodactyle. *Am. Jour. Sci.*, (3), 1, p. 472.

28 Marsh, O. C., 1872. Discovery of a remarkable fossil bird. *Am. Jour. Sci.*, (3), 3, pp. 56–57.

29 Marsh, O. C., 1872. Discovery of additional remains of Pterosauria, with descriptions of two new species. *Am. Jour. Sci.*, (3), 3, pp. 241–248.

30 Hankin, E. H., and Watson, D. M. S., 1914. On the flight of pterodactyls. *Aeronaut. J.*, 18, pp. 324–335.

31 Ibid.

32 Bramwell and Whitfield, noting the resemblance between *Pteranodon*'s feet and those of a bat, infer that *Pteranodon* hung upside down on cliff faces. Bramwell, C. D., and Whitfield, G. R., 1974. Biomechanics of *Pteranodon*. *Phil. Trans. Roy. Soc.*, London 267 B, pp. 503–592. Ostrom (pers. comm. March 4, 1975) points to the lack of detrital material in the Niobrara chalk of Kansas as evidence that any cliffs must have been far removed.

33 This notion, as was shown in Ch. 1, seems to have originated with Richard Owen in 1841 although it was still appearing in the popular press in 1914, see Hankin and Watson, op. cit. (30).

34 Brown, B., 1943. Flying reptiles. *Natural History*, 52, pp. 104–111.

35 Heptonstall, W. B., 1971. An analysis of the flight of the Cretaceous pterodactyl *Pteranodon ingens* (Marsh). *Scott. J. Geol.*, 7 (1), pp. 61–78.

36 Bramwell, C. 1970. The first hot-blooded flappers. *Spectrum*, 69, pp. 12–14. See also Bramwell, C., and Whitfield, G. R., 1970. Flying speed of the largest aerial vertebrate. *Nature*, 225, pp. 660–661. Heptonstall, op. cit. (35), considered *Pteranodon* over twice the weight that Bramwell allowed in her calculations. Using his figures the pterosaur would come in a poor second to the glider in performance. Bramwell and Whitfield have subsequently contested Heptonstall's estimate.

37 Short, G. H., 1914. Wing adjustments of pterodactyls. *Aeronaut. J.*, 18, pp. 336–343.

38 Editorial comment to ibid.

39 Bramwell and Whitfield, op. cit. (32).

40 Heptonstall, op. cit. (35).

41 Hankin and Watson, op. cit. (30).

42 Bramwell, C., 1970. Those flappers again. *Spectrum*, 72, p. 7.

43 Bramwell and Whitfield, op. cit. (32), however, suggest that, being ultra light, the pterosaur carcasses may have floated this far out to sea.

44 Heptonstall, op. cit. (35).

45 Brown, op. cit. (34).

46 See the comment by Short, op. cit. (37), p. 340, on the preceding paper by Hankin and Watson.

47 Halstead, op. cit. (19), p. 142.

48 Lawson, D. A., 1975. Pterosaur from the latest Cretaceous of West Texas: discovery of the largest flying creature. *Science*, 187, pp. 947–948. I should like to thank Mr Lawson for showing me a draft of this paper before its publication. The new Texas pterosaur has not been officially named and the specimen that Lawson wishes to make the type has yet to be excavated. That many popular misconceptions still surround these gliders is apparent from the press statements concerning Lawson's find. 'Announcement of the discovery', claimed *The Times*, 'is expected to rekindle an old debate among palaeontologists over whether flying reptiles flapped their featherless, leathery wings or merely climbed onto high perches and leapt into the air currents to soar like gliders.' *The Times*, March 13, 1975, p. 1. It is apparent that the idea of a hairy pterosaur has yet to be popularly assimilated. See my comments in *The Times*, March 14, 1975, p. 16.

49 Desmond, A. J., 1975. The puzzle over the pterosaur: how did the world's biggest bird[!] ever manage to fly? *The Times*, April 12, p. 12.

8. The coming of Armageddon: a cosmic cataclysm?

1 Reig, O. A., 1970. The Proterosuchia and the early evolution of the archosaurs; an essay about the origin of a major taxon. *Bull. Mus. Comp. Zool.*, Harvard, 139 (5), pp. 229–292. See p. 230.

2 Simpson, G. G., 1953. *Major Features of Evolution.* New York, pp. 291–293.

3 Scars are often visible on the frill, indicating the attachment of muscles in life. The frill also protected the neck region, always the most vulnerable spot. Ostrom, J. H., 1964. A functional analysis of jaw mechanics in the dinosaur *Triceratops. Postilla*, 88, pp. 1–35. See pp. 11–13.

4 Russell, D. A., 1967. A census of dinosaur specimens collected in Western Canada. *Natl. Mus. Can., Hist. Pap.* 36, pp. 1–13.

5 Newell, N. D., 1966. Problems of geochronology. *Proc. Acad. Nat. Sci. Philadelphia*, 118, pp. 63–89. See p. 74. Such a 'revolutionary turnover' in only a million years demonstrates the rapidity with which a whole community might become extinct if there were no successors.

6 Swain, T., 1974. Cold-blooded murder in the Cretaceous. *Spectrum*, 120, pp. 10–12.

7 Colbert, E. H., 1965. *The Age of Reptiles.* London, pp. 161 *et seq.*

8 The tortoise is a very primitive reptile. Its ancestry, even though little understood, is presumed to lie somewhere near the earliest reptiles, soon after their evolution from the amphibians. Both mammals and dinosaurs had far later origins and are equally remote from the tortoise. Since both mammals and dinosaurs were warm-blooded (a factor not considered by Swain: note the title of his paper), there could be reasons for thinking that dinosaurs would be better allied with mammals than cold-blooded tortoises in Swain's study.

9 Thus the British Museum (Natural History) has a display of pathological abnormalities found in dinosaurs which includes a flexed bipedal dinosaur skeleton. Yet, in the nearby *Archaeopteryx* exhibit, casts of the Jurassic bird-dinosaur can be seen showing exactly the same symptoms – without a hint of their having been pathological specimens. There are far too many flexed skeletons for them *all* to be pathological and most authorities now accept that the condition resulted from carcass dessication.

10 Rosenkrantz, A., and Brotzen, F., 1960. International Geological Congress. *Report of the Twenty First Session, Norden.* Part 5. The Cretaceous Tertiary boundary.

11 Hay, W. H., 1960. The Cretaceous-Tertiary boundary in the Tampico Embayment, Mexico. *Ibid.* pp. 70–77.

12 Hall, J. W., and Norton, N. J., 1967. Palynological evidence of floristic change across the Cretaceous-Tertiary boundary in eastern Montana (U.S.A.). *Palaeogeography, Palaeoclimatol., Palaeoecol.*, 3, pp. 121–131. In the late Mesozoic conifers accounted for only 12% of the flora, but by the early Tertiary they had increased their hold to 30%. In absolute numbers, the authors found 11 species of gymnosperms in the Hell Creek formation (latest Cretaceous) and 16 species in the Tullock formation (earliest Tertiary). Dicotyledons (flowering plants) ran at 46 species in Hell Creek but only 17 in Tullock. The total number of plant species was halved across the boundary.

13 Axelrod, D. I., and Bailey, H. P., 1968. Cretaceous dinosaur extinction. *Evolution*, 22, pp. 595–611. *Metasequoia* and other 'living fossils', like the ginko or maiden-hair tree cultivated in Chinese and Japanese temples, have found a refuge in our world in some upland areas in China, where they are associated with hardy evergreens and deciduous hardwoods. At lower altitudes they give way to lush forests of oak, palm and laurel.

14 Bramlette, M. N., 1965. Massive extinctions in biota at the end of Mesozoic time. *Science*, 148, pp. 1696–1699.

15 Terry, K. D., and Tucker, W. H., 1968. Biologic effects of supernovae. *Science*, 159, pp. 421–423. These and subsequent figures in the text are taken from this source. There is a measure of independent evidence that direct solar radiation (but of a smaller quantity) may cause faunal change, at least among unicellular organisms. Every million years or so the earth's magnetic field changes direction, so the north pole becomes south and *vice versa*. When the flip is about to occur the intensity of the field decreases to about 15%, the switch occurs in a thousand years, and the field builds up strength again. During the flip the van Allen belt, which protects the earth from cosmic radiation, collapses and the earth is bathed in solar radiation. There is a remarkable degree of correlation between these magnetic flips and the vicissitudes of radiolarian populations. As the poles switch, the radiolarians – single-celled organisms about 0.1 mm in size and with a hard shell – concomitantly change their characteristic type. Uffen, R. J., 1963. Influence of the earth's core on the origin and evolution of life. *Nature*, 198, pp. 143–144. See also: Black, D. I., 1971. Polarity reversal and faunal extinction. *In* Gass, I. G., Smith, P. J. and Wilson, R. C. L. (eds). *Understanding the Earth.* Artemis, pp. 257–261.

16 Laster, H., 1968. Cosmic rays from nearby supernovae: biologic effects. *Science*, 160, p. 1138.

17 Russell, D., and Tucker, W., 1971. Supernovae and the extinction of the dinosaurs. *Nature*, 229, pp. 553–554.

18 Russell, D., 1971. The disappearance of the dinosaurs. *Canad. Geog. J.*, 83, pp. 204–215.

19 Clemens, W. A., 1960. Stratigraphy of the Type Lance formation. Rosenkrantz and Brotzen, op. cit. (9), pp. 7–13.

20 Colbert, op. cit. (6), p. 200.

21 Ibid. pp. 197–203. See also pp. 169–179.

22 Between the Cenomanian (opening of Upper Cretaceous) and lower Maestrichtian the number of ammonite families recorded dropped from 22 (containing 78 genera) to 11 (with 34 genera). During the Maestrichtian only 11 of these genera managed to survive to the end when they were all destroyed. There is also a notable contraction of ammonite range; at the opening of the Maestrichtian ammonites were world wide, but towards the close they had already vanished from the southern hemisphere. Essentially the same story can be told of the related belemnites. Hancock, J. M., 1967. Some Cretaceous-Tertiary marine faunal changes. *In* Harland, W. B., *et al.* (eds). *The Fossil Record*. Geological Society, London, pp. 91–104.

23 Cys, J. M., 1967. On the inability of the dinosaurs to hibernate as a possible key factor in their extinction. *J. Paleontol.*, 41, p. 226.

24 Bakker, R. T., 1972. Anatomical and ecological evidence of endothermy in dinosaurs. *Nature*, 238, pp. 81–85. Cys, op. cit. (22), estimated that the smallest dinosaurs in the uppermost Cretaceous, namely *Leptoceratops*, *Stegoceras*, *Dromaeosaurus*, *Saurornithoides* and *Thescelosaurus*, were no less than 5 to 12 feet long as adults.

25 Bakker, R. T., 1971. Dinosaur physiology and the origin of mammals. *Evolution*, 25 (4), pp. 636–658. Large hatchling size is considered on p. 651.

26 Russell, L. S., 1965. Body temperature of dinosaurs and its relationships to their extinction. *J. Paleontol.*, 39, pp. 497–501. In this paper Russell tentatively explores the consequences that would have followed if the dinosaurs had been warm-blooded. He concluded that endothermy was 'possible' and indeed 'plausible'. Naked skin in dinosaurs, especially if the beasts were endothermic, would have resulted in a certain susceptibility to the cold. Published as early as 1965, Russell's short but seminal paper is full of ideas amplified and developed by later writers. Naked skin is also treated in Bakker, op. cit. (24), p. 650.

27 Bramwell, C. D., and Whitfield, G. R., 1974. Biomechanics of *Pteranodon*. *Phil. Trans. Roy. Soc.*, London, 268 B, pp. 503–592.

28 Russell, D. A., 1975. Reptilian diversity and the Cretaceous-Tertiary transition in North America. *In* Caldwell, W. G. E. (ed.). Colloquium on the Cretaceous System in the Western Interior of North America. *Geological Association of Canada*, *Special Paper*, (in press).

29 Reported in Russell, D. A., 1973. The environments of Canadian dinosaurs. *Canad. Geog. J.*, 87(1), pp. 4–11.

30 Erben, H. K., 1972. Ultrastrukturen und Dicke der Wand pathologischer Eischalen. *Abh. Akad. Wiss. Lit, math.-nat. Kl.*, 6, pp. 191–216. See also Tafel I, Figs. 5 and 6.

31 Professor Erben informs me that the decrease in shell thickness was *not* sudden but a gradual affair. Nevertheless, he is of the opinion that the whole process took place *very* quickly. Erben's evidence suggests that the stress on the dinosaurs did not appear with 'supernova' abruptness, even though the time involved was rapid by geological standards. Pers. comm. 27 Feb., 1975.

Appendix 2

1 Popper, K. R., 1972. *Conjectures and Refutations*. London, p. vii.
2 Feduccia, A., 1973. Dinosaurs as reptiles. *Evolution*, 27, pp. 166–169. Dr. Pamela Robinson takes a similar (if extreme) view with regard to the acquisition of warm-bloodedness in mammals. She argues that mammalian endothermy probably evolved only in the late Mesozoic or early Tertiary, concomitantly with the enlargement of the mammalian brain. Robinson, P. L., 1971. A problem of faunal replacement on Permo-Triassic continents. *Palaeontology*, 14, pp. 131–153.
3 Russell, D. A., 1969. A new specimen of *Stenonychosaurus* from the Oldman Formation (Cretaceous) of Alberta. *Can. J. Earth Sci.*, 6, pp. 595–612. Russell, D. A., 1972. Ostrich dinosaurs from the late Cretaceous of Western Canada. Ibid., 9, pp. 375–402.
4 Feduccia, A., 1974. Endothermy, dinosaurs and *Archaeopteryx*. *Evolution*, 28, pp. 503–504.
5 Jerison, H. J., 1973. *Evolution of the Brain and Intelligence*. New York and London, p. 408.
6 Bakker, R. T., 1971. Dinosaur physiology and the origin of mammals. *Evolution*, 25, pp. 636–658.

Credits for illustrations

Figures 1,2	Faujas-Saint-Fond, B., 1799. *Histoire Naturelle de la Montagne de Saint-Pierre de Maestricht*. Paris. By permission of the Trustees of the British Museum.
3	Osborn, H. F., 1899. *Science*, 10, pp. 919–925.
4,5,36	*Illustrated London News*: 1853, 23, pp. 599–600; 1854, 24, p. 22; Restoration by Neave Parker, 1955, 227, p. 1065.
6,7	Photographs taken specially for the book by Keith Penny.
8	Left: Owen, R., 1854. *Geology and Inhabitants of the Ancient World*. London. Right: Colbert, E. H., 1968. *Men and Dinosaurs*. London. Original source unknown.
9	Restoration by Charles R. Knight, *in* Osborn, H. F., 1931. *Cope: Master Naturalist*. Princeton.
10	*Harper's Weekly*, 1869, 13, p. 525.
11	Board of Commissioners of the Central Park, 1868. *Twelfth Annual Report*. New York.
12	Hitchcock, E., 1836. *Am. Jour. Sci.*, 29, pp. 307–340.
13,29,49,50,51	Heilmann, G., 1926. *The Origin of Birds*. London.
14	Osborn, H. F., 1911. *Amer. Mus. Jour.*, 11, pp. 7–11.
15	Lambe, L., 1920. *Canada Dept. Mines, Geol. Surv.*, 120, pp. 1–79.
16	Restoration by Robert T. Bakker, *in* Russell, D. A., 1973. *Canad. Geog. J.*, 87, pp. 4–11.
17,20,22	Matthew, W. D., 1915. *Dinosaurs*. New York: American Museum of Natural History.
18,19,32,37,63,64	Prepared specially for the book by Jessica Gwynne.
21,52,60	Augusta, J., and Burian, Z., 1961. *Prehistoric Reptiles and Birds*. London.
23	Bird, R. T., 1954. *National Geographic*, 105, pp. 707–722.
24,31,34,48	Restorations by Robert T. Bakker, *in* Crompton, A. W., 1968. *Optima*, 18, pp. 137–151.
25,26	Restorations by Robert T. Bakker, *in* Ostrom, J. H., 1969. *Bull. Peabody Museum of Natural History, Yale University*, 30, pp. 1–165.
27,28(left)	Galton, P. M., 1970. *J. Paleo.*, 44, pp. 464–473.
28(right)	Russell, D. A., 1972. *Can. J. Earth Sci.*, 9, pp. 375–402.
30	Restoration by Robert T. Bakker, *in* Galton, P. M., 1974. *Bull. Br. Mus. nat. Hist. (Geol.)*, 25(1), pp. 1–152.
33	Romer, A. S., 1971. *Breviora*, 378, pp. 1–10.
35	Kermack, K. A., 1965. *Science Journal*, 1(9), pp. 66–72.
38	Phillips, J., 1871. *Geology of Oxford and the Valley of the Thames*. Oxford.
39	Ballou, W. H., 1897. *The Century Illustrated Monthly Magazine*, New York, 55, pp. 15–23.
40	Holland, W. J., 1905. *Mem. Carnegie Mus.*, 2, pp. 225–279.
41	Anonymous [Holland, W. J.], 1905. *Ann. Carnegie Mus.*, 3, pp. 443–452.
42	Hay, O. P., 1910. *Proc. Wash. Acad. Sci.*, 12, pp. 1–25.
43,44	Holland, W. J., 1910. *Am. Nat.*, 44, pp. 259–283.
45	Holland, W. J., 1924. *Mem. Carnegie Mus.*, 9, pp. 379–403.
46	Osborn, H. F., and Mook, C. C., 1919. *Am. Philos. Soc.*, 58, pp. 386–396.
47	Bird, R. T., 1944. *Natural History*, 53(2), pp. 60–67.
53	Marsh, O. C., 1880. *Odontornithes*. Washington.
54	Matthew, W. D., and Granger, W., 1917. *Bull. Amer. Mus. Nat. Hist.*, 37, pp. 307–326.
55	Abel, O., 1911. *Die Abstammungslehre*, pp. 198–230.

56 Hawkins, T., 1840. *The Book of the Great Sea-Dragons*. London.
57 Phillips, S., 1856. *Guide to the Crystal Palace and Park*. London, p. 190.
58 Broili, F., 1938. *Sitzgsb. d. math.-naturw. Abt.*, pp. 139–154.
59 Edinger, T., 1941. *Am. Jour. Sci.*, 239, pp. 665–682.
61 Newman, E., 1843. *The Zoologist*, 1, pp. 129–131.
62 Eaton, G. F., 1910. *Mem. Connecticut Acad. Sci.*, 2, pp. 1–38.
65 Osborn, H. F., 1917. *Bull. Am. Mus. Nat. Hist.*, 35, pp. 733–771.

Acknowledgements

I should like to extend my thanks to John H. Ostrom, who read the completed manuscript and added many valuable comments and criticisms. Frances Mussett corrected Chapters 4 and 7, and R. B. Freeman and Dale A. Russell checked Chapters 1 and 8 respectively. The following also contributed, either by way of conversation or correspondence: Donald Baird, Heinrich K. Erben, Kenneth A. Kermack, Douglas A. Lawson, Pamela L. Robinson, Charles Coleman Sellers and Tony Swain. Barbara Desmond and Jaqueline A. Cowie read chapters. It is a pleasure to thank Pamela Gill, who not only checked the many drafts of the manuscript, but contributed immeasurably to the text with ideas and suggestions.

Jack Saxon, Leo Desmond and Geraldine Stanley assisted with the Russian, Italian and German translations. Jessica Gwynne was responsible for the new flesh restorations, and Keith Penny took the photographs of the Crystal Palace reconstructions. Eva Crawley photographed the text-illustrations.

The research was carried out in the D. M. S. Watson Natural Sciences Library at University College, London, using the extensive reprint collections of the late D. M. S. Watson and Arthur Smith Woodward. Additional material was gathered from the libraries of the British Museum (Natural History), Senate House (London University) and the Museum of Comparative Zoology at Harvard University. I am indebted to the librarians of University College, London for tolerating my unremitting demands. Finally, my thanks go to Anthony Blond and Desmond Briggs for their constant encouragement.

Finally, I wish to express my gratitude to Everett Mendelsohn, Stephen Jay Gould, and the Faculty of the Department of the History of Science, Harvard University, for granting me a leave of absence for 1974–5, during which time this book was in part conceived and written.

Index

232

233